I0462070

Published By

Those Who Truly Care

Ghana, West Africa

2009 Here After

Dedicated to:

"The Pure Ones"

Living In Spirit and Truth

"Guided By the Living Forces of Almighty Eternal Nature"

The Symbol of Solar Cycle of RE (RA)

3x3=9

The Three triads

1. The Mental, the Emotional, the Vital Principle.
2. The Intelligence, the intuition, the will
3. The Creator, the Preserver, the Transformer

"All aspects reflected in us Deity, coming from RE the Supreme"

-NU-

(NUN)

Ancient (**TaMa-Rean**) Egyptian Word and Principle for 1st, Prime, **Hidden Potential** and **Infinite Possibilities**. The Original Creative Forces of the Universe (Nature), the **Black Light** of **Intellect** (*Mental Energy*), the **Black Waters** or **Celestial Waters** known as the **Etheric Realm**, formless **Ethers**, Un-manifested energies, Nameless.
The Eternal Now!

*"As **African People** our **Minds** are linked to **ALL Eternal** and **Almighty Nature**, what we can conceive in our **Minds** will be, see it and so shall it be seen".* – ***Neb Heru***

"*YOU*" –*The Real Student of Life!*

"With the **Ankh** "**Master Key**" in Her or His possession, the **Initiate** may unlock the many doors of the **Mental** and **Psychic Temple of ALL**, thus gaining **The Right Knowledge, Right Wisdom** and **The Right Overstanding**, and enter the same freely, using the **GOD Mind** you have (**As the Children of Nature and the Initiates of Life**) or the **Mind of Mental**, which is **GOD**. This explains the **true Nature of Energy, Power, Matter, Existence**, and why and how all these are subordinate to the **Mastery of Mind**, and that each **Mind** is a **slave to the Mental, the Force of Ether**, which controls the **Action of Matter**."

One of the old *Tehuti Masters* wrote, long ages ago: "He or She realizes that "**ALL**" is, and these words are as true for all time. Without this **Ankh** "**Master Key**", mastery is impossible, and the **Initiate** knocks in vain at the many doors of the **Temple**, and is met by **false Teachers of Ghost, Illusions, Spooks, Fictions, Beliefs**, and **Myths, given false hopes. No Facts, No Confirmations, No Truth**, just lie after lie. The Light is devil's (**Six Ether Death Forces in Nature**) tool. **Darkness** is the home (**Abode**) of **GOD** (**Deity, Neteru**) **Peace** (**Hotep, Hetep**), **Tranquility**, and **Bliss**.

What is Qatum (Melanin) Physics?

In this day and time the Students (**Initiates**) of **Life (Nine Ether)** must find your way back home to **Blackness (NUN) the State of Supreme Mind**.

*"The **Sacred Wisdom** of the **Grand Hierophant Tehuti**"*

Diagram 1
Tehuti holding the "Nu Bowls"
Frankincense and Myrrh offerings

ANKH - The Master Key

Question: What is the Master Key?

Answer: The Master Key is based on absolute Scientific Laws of Nature (**NUN**), Facts, and will unfold the possibilities that lie dormant in **each individual**, and teach how they may be brought into powerful action, to increase the person's effective capacity, bringing added energy, "**Conscience**" the ability to discern from what is right from that which is ultimately wrong in one's life. The Person possessing the **Master Key** will find vigor, a renewed excitement for life and mental elasticity. The Person who gains an overstanding of these **Natural Mental Laws of Nature** which are unfolding will come into the possession of an ability to secure results

undreamed of, and which has rewards hardly expressed in words. You as **Ancient Egyptians** have finally arrived where you can take your place as **Divine Rulers**, of **PTAH-NUN** (Earth), **PAA RE** (THE SUN) and **PAA PAUT** (THE ALL).

With the use of **the Master Key** you will learn the use of "**Mind Power**", which is true **God Power, Ruler Ship**, which is **Deity Ship**. Re-learning how to use the hidden (**AMUN**) powers of your **Mind** has nothing to do with **Magic** or **Hypnotism**, even though to the unlearnt these concepts where birth. When we are back in our right state of **Mind Power**, what seems like **Magic** and **Miracles** to the unlearned and unknowing on looker, will be simple **<u>Fundamental Laws of Nature.</u>**

Possessing the **Master Key** allows the Person to cultivate and develop the overstanding which will enable the Person to control the body and thereby the Health. Having the **Master Key** in your possession improves and strengthens the **Mind** and **Memory**. It develops insight and foresight, the type of insight and foresight that is so rare, the kind which is the distinguishing characteristics of every successful, and "**Powerful and Great Ancestor**" of the past.

The Master Key develops **Mind Power** which means that others instinctively recognize that you are a person of force, of character – that they want to do what you want them to do in the "**Positive**"; it means that you "**attract**" people and things to you; that you are what some people call "**lucky**", that "**things**" come your way with easy, that you have come into the an <u>overstanding</u> of the <u>Fundamental Laws of Nature</u>, and have put yourself in harmony with them; that you are intune with **Infinite Nature**; that you **Overstand the Law of Attraction**, <u>the Natural laws of Growth and Change</u>, and the <u>Inner Laws of Nature</u> on which all advantages in health, Social and Business World rest.

Mental Power (<u>The Master Key</u>) is creative power which is **GOD Power**, gives you the ability to create for yourself; it does not mean to take something away from someone else. Nature never does things that way. Nature the Ultimate provider makes two blades of grass grow where one grew before, and **Mind Power** enables people to do the same thing.

With the **Master Key** in your possession you will see that it helps you to develop insight, increased independence, and the ability to be helpful. It destroys, fear, distrust, depression, melancholia, and every lack, limitation and weakness, including pain and disease. It **awakens buried talents, supplies initiative force**, **energy**, and **vitality**. **The Master Key** awakens an appreciation of the **beautiful** in your own **race**, **self pride** and a sense of **inner greatness**. You will learn to rise, not fall back in love with self and kind again, and have a greater appreciation for your great Ancient Culture and the great minds of your **Ancestors**.

The Master Key has changed the lives of all the **Great Men** and **Women** of today and of our Great Ancient Past, the True Egyptian know today as **Nuwaupian**. This is truly our day and time! Once you learn to substitute definite **Principles of Nature** which is the **Laws in Nature**, the facts, for beliefs, hazy methods, and principles for the foundation and **Fundamental Laws of Nature**, a great inner strength will arise. You will have facts beyond a shadow of a doubt that your **Ancient Ancestral Forces of Nature** are with you, in you, and working for you in your everyday life.

Having the **Master Key** you will learn <u>**Sound Right Reasoning**</u>, in ancient **TaMa-Re (Egypt)** called **Nun**. **The Master Key** teaches **Right Principles**, and suggests methods for making "**Practical Application**" of the **Principles**. With the **Master Key** in your possession you the true student of **Ancient Egypt**, you will learn the value of "**Application**", you will now sit amongst the **Doers**, the **Knower's**, and those who **Reason**. With the **Master Key**, the true Student of **Egypt**, will learn the value of putting to practice **the Eternal and Almighty Laws of Nature**, with this you see **True Facts** manifest not spooky beliefs.

In this new Millennium, Science has taken on new leaps, and bounds. They have tapped into more **facts** about the vast universe and its infinite possibilities. In this new Millennium called the *Solar Cycle of RE* (**RA**) People are learning the value of **the Mind**, and **the Mental Universe**, we are all learning **Mind over Matter.** Grab your **Master Key** and come up here and join us true **Mental Giants.** The Whole World is on an Eve (**Neith – Biaps the Mother of Energy, Mitochondria DNA**) of a new (**NU or NUN**) Consciousness.

"Mine your Mind for the Jewels of your Soul!"

ARE <u>YOU</u> IN THERE?

If yes

Let's Start!

TABLE OF CONTENTS

Question: What is Qatum Physics?

Answer: The Word **Qatum** coming from Ancient Egyptian Mystery Language called Nuwaupic meaning "<u>Melanin</u>" and according to the Merriam Webster's Collegiate Dictionary the word "**Physics**" mean: *Greek - Physika,* Plural *Physikos* of **Nature**. Fr. *Physis* **Growth, Nature**. From *Phyein* **To Bring Forth**. "**A Science that deals with <u>Matter</u> and <u>Energy</u> and their interactions**".

Question: Is there any relation between the word Qatum in Nuwaupic and Quantum in English they sound the same?

Answer: Yes, the reason why we say **Qatum (Melanin)** in our **Ancient Egyptian Mystery** Language called **Nuwaupic** and the word "**Quantum**" are the same words phonetically, because as you know we **Ancient TaMa-Rean (Egyptians)** known today as **Nuwaupians**, saw ourselves as the **Supreme Beings, Children of Nature (Neteru)**, Children of **PAUT (ALL)** as a **Quantum Being**. **Quantum** meaning "**four**" representing the **4 principles** and **Cycles in Nature**, in **TaMa-Re (Egypt)** they where know as **Atum-Re, Atun-Re, Amun-Re** and **Anun-Re**, giving you a **Quadity Principle**, and not a **Trinity**.

In ancient **TaMa-Re** we always acknowledged the **Female (Feline) Forces** of **Nature** or the **Feminine Force** or **Principles** in **Nature**. We **Africans**, represent a mighty race of People that have survived all Cycle changes within **Nature**, linking us to **PAUT** (ALL). We Ancient **TAMA-REANS** Children of **TA** (Earth), **MA** (Water), and **RE** (SUN), acknowledge ourselves as **Quantum Beings**, because of our physical make up. Within our skin we have Gold and that Gold we refer to is "**Melanin**". **Melanin**, manifest on this side of **Hydrogen** (H1) from the **Black Etheric** womb of the **Higher Planes** or **Abodes of Existence** as **Light**, and when you see **Full Spectrum Light** in its **Physical State**, it manifests as **Blackness**, and the Manifestation of the **Solid form of Light** is **Melanin**.

As **Qatum Beings** and **Children of PAA RE (The Sun)** what we must start to acknowledge and become aware of is that everything is connected and is defined as different **Rates**, or **Modes of Vibration**. So through your **DNA** and **Melanin** and your whole physiological (body) make up as a **Melaninite, 9 Etheric Being** you are linked to the **Universe**, linked to the **Sun (PAA RA, RE)**, **Planet Earth (PTAH-NUN)** and all other **Celestial Bodies**. **Melanin** or what people are calling **Melanin** is really **Carbon** based, also know in our **NUN Science** as **Sun Heat Genes.**

Sun Heat Genes are in actuality a burning that is taking place in the cells, which are known as **Mast Cells,** and that burning that is taking place actually burns towards the service of the skin. Remember <u>life is a burning</u>! When you take the time to look at Male, **Masculine Energy** with a positive (+) charge, *"The Giver",* and Female, **Feminine Energy** with a negative charge (-) *"The Receiver"* come together and cause friction or Heat it causes a <u>**Spark of Light**</u> which is the 1st point of Light (**Atum-Re**). **The Sperm** (the Light or Life carrier) travels up the spine of the Male to the Crown Seat (**Pineal Gland**) and receives the 2nd spark (**Atun-Re**), this newly charged Sperm cell travels back down through the **Uretha in Males** and gushes forth into **the deep primordial abyss,** *"The Womb"* and reaches the egg and the 3rd point of light occurs inside the womb (**Amun-Re**). From **Amun –Re** to **Anun-Re** to **Atum-Re, 3 points of Darkness,** all happening within the **Principle of the Quadity,** and linking our being to **Quantum** (**The ALL in ALL**). The Minds of the world are now ready to receive **Quantum Physics, Quantum Reality, which is the level of Full Conscious Awareness.**

What is Qatum (Melanin) Physics?

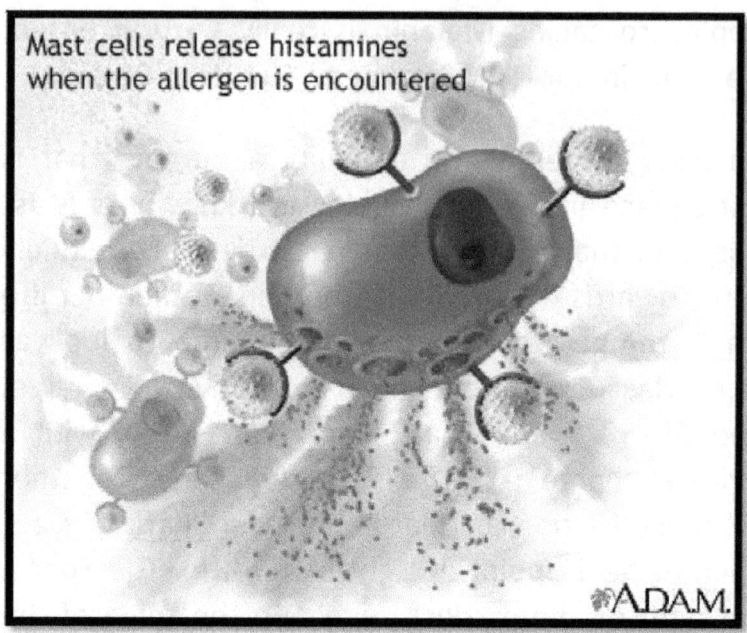

Mast cells release histamines when the allergen is encountered

MAST CELLS - The Home of Sun Heat Genes

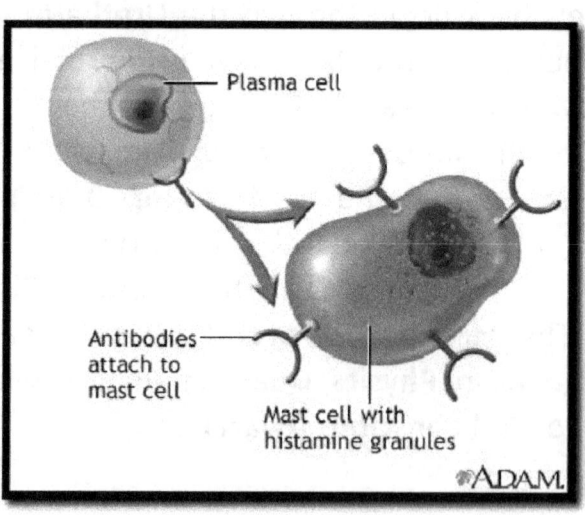

Plasma cell

Antibodies attach to mast cell

Mast cell with histamine granules

Question: Is there a connection between the word Qatum for Melanin in Our Ancient Egyptian Mystery Language called Nuwaupic and Quantum Physics?

Answer: Yes there is definitely a connection, through your **Melanin** which is a condensed form of **Light**, you as a **Qatum** being are connected to **ALL**, that which Scientist are calling **Quantum**.

Melanin is a pigment responsible for the color of skin. When **Melanin** does not appear in the skin, it is deficient; however this has nothing to do with the amount in your **Brain** called **Neuromelanin** also know as **Supreme Melanin**.

Neuromelanin is a form of **Melanin** found in the **Central Nervous System.** Scientist and doctors today try to reduce the importance of **Melanin**, by overlooking the significant difference between **Brain Melanin** and **Skin Melanin** properties. **Neuromelanin** does not run parallel with your **Skin Melanin**. Whether white, red, yellow, black, brown, or albino, **Neuromelanin** plays an important role in functioning of the **Brain** and **Nervous System**.

What is Qatum (Melanin) Physics?

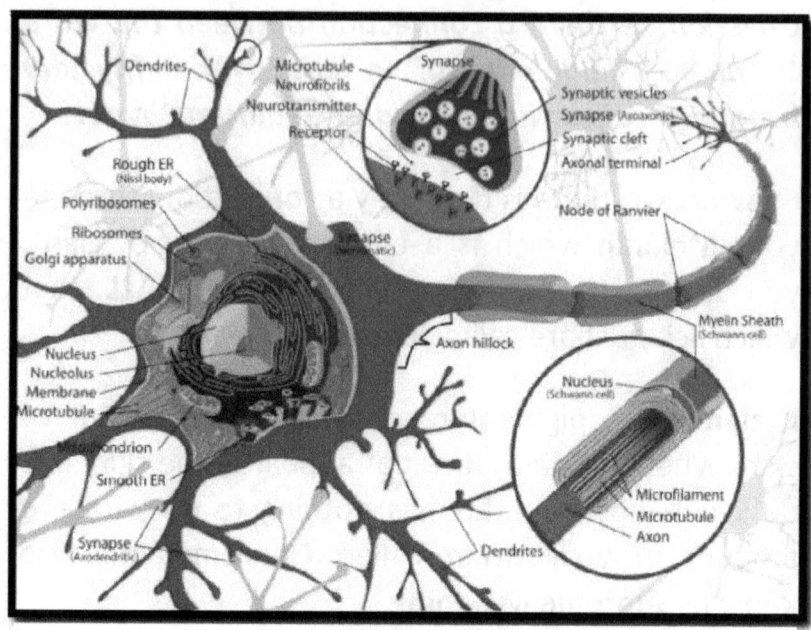

A Brain Neuron

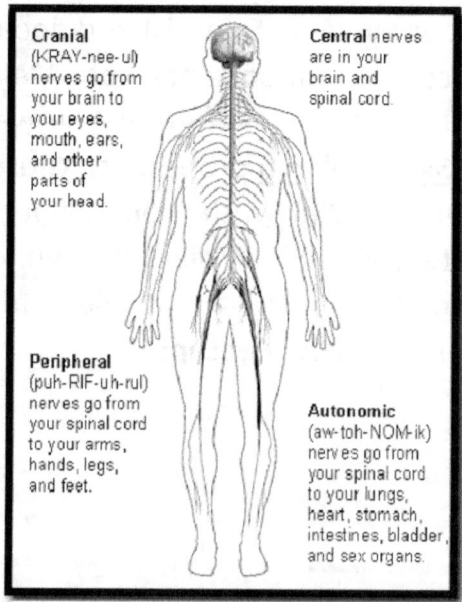

Cranial (KRAY-nee-ul) nerves go from your brain to your eyes, mouth, ears, and other parts of your head.

Central nerves are in your brain and spinal cord.

Peripheral (puh-RIF-uh-rul) nerves go from your spinal cord to your arms, hands, legs, and feet.

Autonomic (aw-toh-NOM-ik) nerves go from your spinal cord to your lungs, heart, stomach, intestines, bladder, and sex organs.

CENTRAL NERVOUS SYSTEM

Question: What is the importance of the Nervous System?

Answer: The **Nervous System** is the **Communications** and **Control Network** of the **Body**. Through **Electrochemical** (**Electric Chemical**) impulses it senses, sorts, stores and uses **messages** that control everybody function and process. It is also responsible for thought, strategy and decision making. Your **Nervous System** is composed of an enormous **Network of Nerve Fibers** and cells that reach every part of the Body.

The **Cerebrospinal Nervous System** or **Central Nervous System** consists of the **Brain** and the **Spinal cord**. It coordinates almost all the activities of the body. The **Central Nervous System** receives information from all parts of the **Body** by incoming **Sensory Nerves**; It then sorts out this **information** and it sends this **information** in its proper orders through the outgoing motor nerves and then to the parts of the body requiring adjustment.

NERVOUS SYSTEM

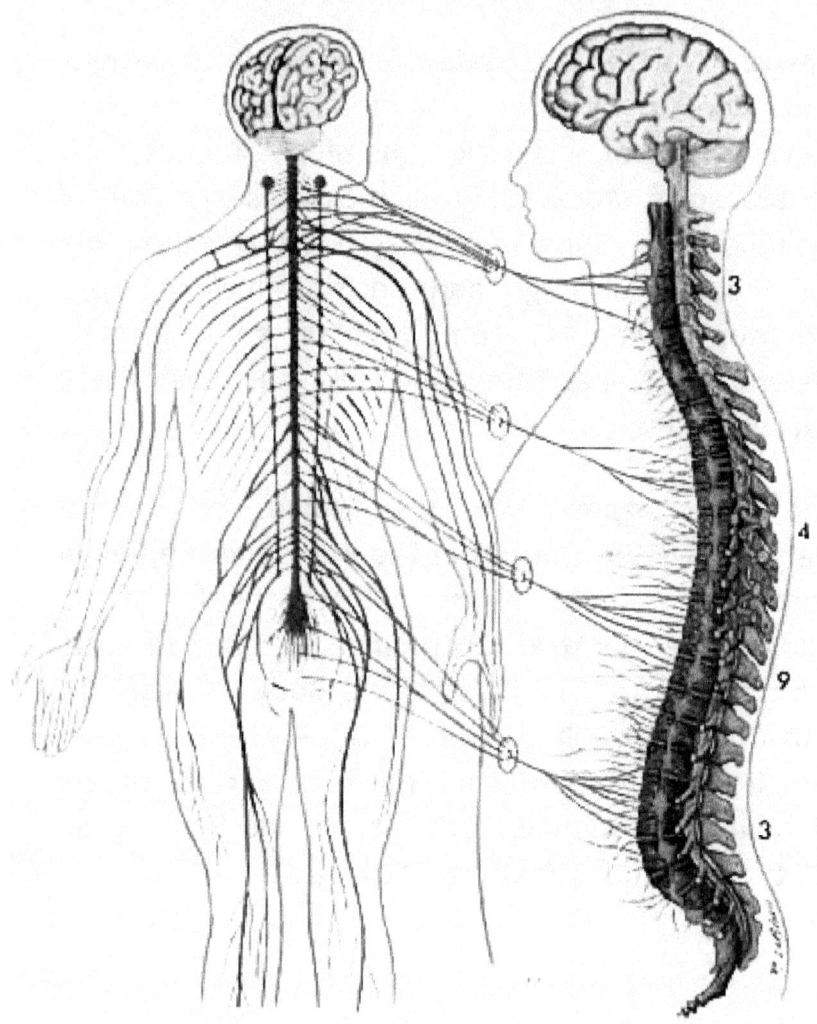

COPYRIGHT 1994 BY CONCEPT THERAPY INSTITUTE, INC. • SAN ANTONIO, TEXAS • 1-800-531-0628

The Nervous System - Electrical System of the Body

Melanin - A Gift from Nature (Neteru)

It is a known fact that **Melanin** is found in abundance in **African People** all over the world. **Melanin** is a gift from the **higher Forces of Nature**, in **Ancient Egipt** called **Neteru**.

Question: What is Melanin?

The **Pigment Melanin** is produced by **cells** in the **Epidermis** called **Melanocytes**. **Melanin** is defined as: *According to the Merriam Webster's Colligate Dictionary* as: ***Melanism** – 1. An increased amount of **black** or nearly **black pigmentation** (as of skin, feathers, or hair) of an individual or kind of organism. 2. Intense Pigmentation in Man in skin, eyes, and hair – **Melanistic,** Adj.

Melanin – 1. Any of various black, dark brown, reddish brown, yellow pigments of animal or plant structure (as skin or hair)

Melanocyte – 1. An **Epidermal cell** capable of Synthesizing **Melanin**.

Melanocyte Stimulating Hormone – 1. A Hormone secreted by the Pituitary Gland that regulates skin color in Human Beings and other

What is Qatum (Melanin) Physics?

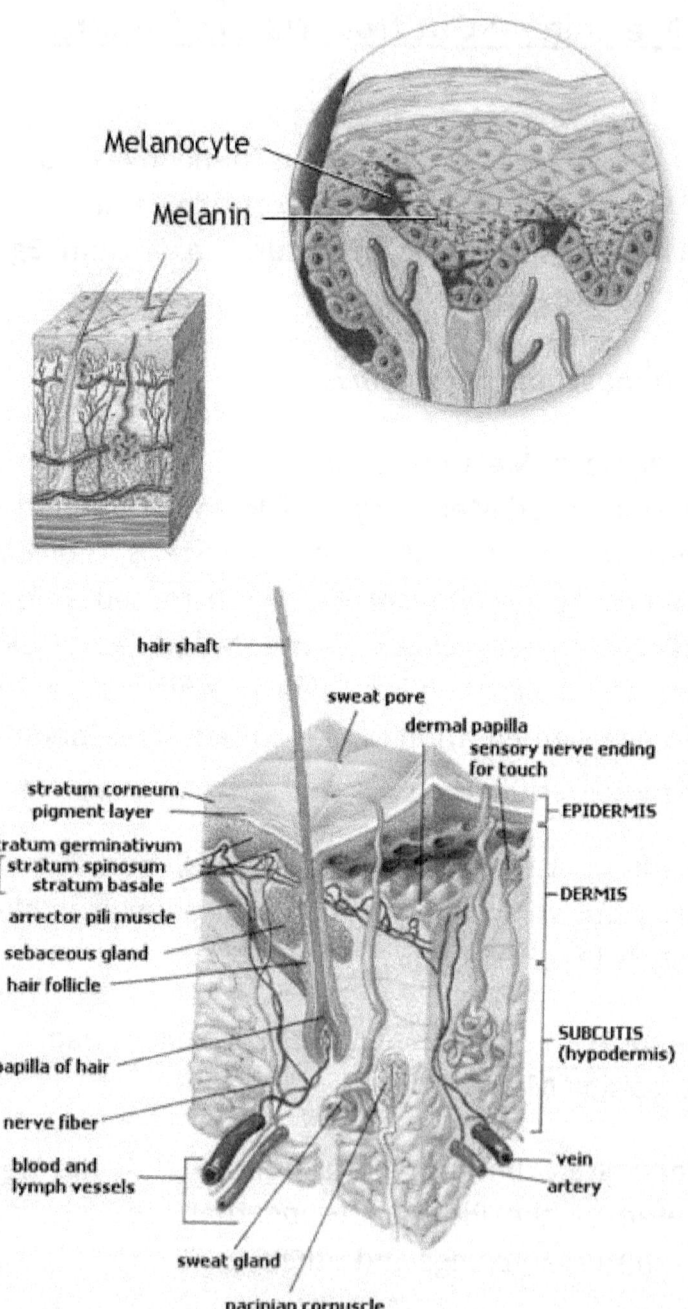

Melanocyte

Melanin

hair shaft

sweat pore

dermal papilla

sensory nerve ending for touch

stratum corneum
pigment layer

EPIDERMIS

stratum germinativum
  stratum spinosum
    stratum basale

DERMIS

arrector pili muscle

sebaceous gland

hair follicle

papilla of hair

SUBCUTIS
(hypodermis)

nerve fiber

blood and
lymph vessels

vein

artery

sweat gland

pacinian corpuscle

Vertebrates by stimulating **Melanin** synthesis in **Melanocytes** and **Melanin granule** dispersal in **Melanophores** also called **Intermedin.**

Question: Why is Melanin important?

Answer: Excellent question! **Melanin** is important, because it is the most **Primitive meaning Prime, Primary – 1st**, and **Universal Pigment** in **Living Organisms. Melanin** is produced in the **Pineal Gland.** Abundantly found in **Primitive Organisms** such as **Fungi**, as well as advanced **Primates.** Furthermore, within each living organism, **Melanin** appears to be located in major functional sites. For example, in Vertebrates, **Melanin** is not only present in the skin, eyes, ears, **Central Nervous System**, and the diffuse **Neurodocrine Loci**, among other things.

Melanin can be found in the **Pineal Gland (The Black Dot), Pituitary Gland, Thyroid Gland, Thymus Gland, Parathyroid, Adrenal Gland,** and **Carotid Body. Melanin** is abundantly present in the viscera, including the **Heart**. The **Liver**, the **Arteries**, the **Muscles**, the **Gastrointestinal** tract, and the **Gonads.** Thus within each and every living organ which aids the Human body **Melanin** appears. Regardless of what color your skin appears to be all **Genes** in all creatures on this planet are black they are all coated with **Melanin.**

What is Qatum (Melanin) Physics?

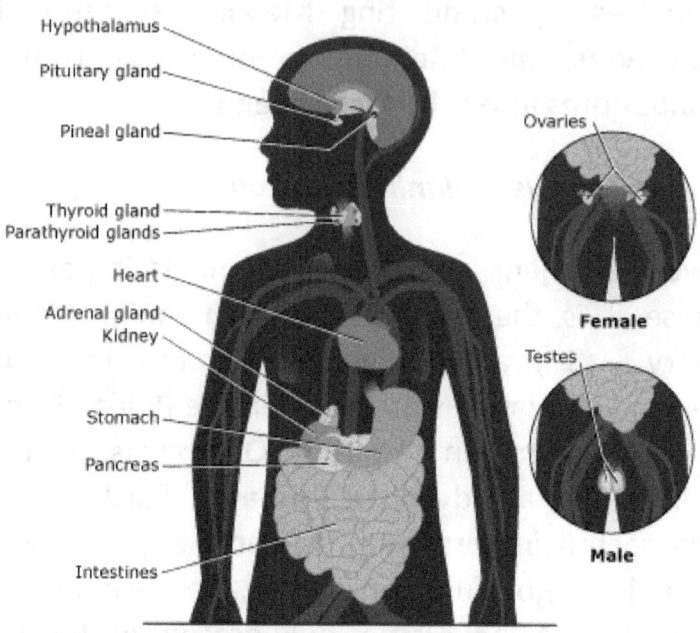

The Endocrine System

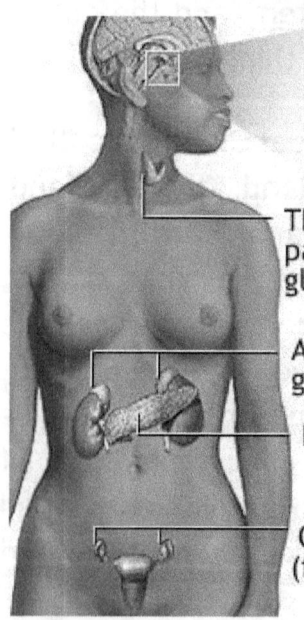

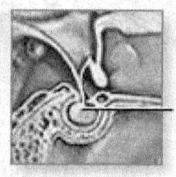

Pituitary gland

Thyroid and parathyroid glands

The endocrine glands secrete hormones which regulate various functions throughout the body

Adrenal glands

Pancreas

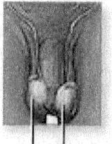

Ovaries (female)

Testes (male)

ADAM.

Question: How much Melanin can be in the Skin of any one individual?

Answer: The amount of **Melanin** in the skin is one of the most variable of Human traits, and many Polygenes' are involved. Groups of people or the population of the world were once classified according to skin shade: Black, (Africans) White (Caucasians – Europeans), Yellow (Orientals) and Red (Indians) Etc... You should realize that just because this is the way they have classified people does not mean this is a fact. The Hues of color of your skin depends on several factors. First is the amount of **Melanin** in the outer layers of the skin. **Melanin** acts as a filter to prevent damage to the delicate deeper layers of the skin, by penetration of **Ultraviolet Light (Black Light)**.

Question: Is there only one Type of Melanin?

Answer: No, of course not! As mentioned earlier you have **Brain Melanin**, known as **Neuromelanin** called "**Supreme Melanin**" which must be activated in this day and time, and then you have **Skin Melanin**. **Neuromelanin** does not run parallel with skin melanin. No matter what race, **Neuromelanin** plays an important role in the functioning of the **Brain**, and **Nervous System**. **Melanosomes** (Small structures within the **Melanocyte Cells** where **Melanin** is synthesized) find their way into the hair cells, giving them color. (Two types of **Melanin**, one dark brown and one red, are responsible for all hair shades.)

Pigments that contribute to skin color are called **Carotene** (Yellowish), **Hemoglobin**, in blood vessels (Pink-Red), and **Melanin** (Black, Brown, and Red). Darker skins are dominated by **Melanin**, which is produced from **Amino Acid Tyrosine**, by Pigment Cells (**Melanocytes**) in the Skin.

Question: Can you explain again how skin color is determined?

Answer: Sure, **Melanin** is produced from the **Amino Acid Tyrosine**, by pigment cells called **Melanocytes**. You have the sacs within the **Melanocyte**, called **Melanosomes** that actually make the **Melanin**. These little sacs or chambers of **Melanin**, determine your skin color. When there is a little bit of **Melanin** in the **Melanosome**, the little sac or chamber, you look white (pale). When it's totally filled you look Black. Skin tone is determined by amount and type of **Melanin**, the **Melanosomes** contains.

What is Qatum (Melanin) Physics?

Amino acids	Letter code		Residue mass (in Daltons)	
Name	Single letter	Three letter	Monoisotopic	Average
Glycine	G	Gly	57.02146	57.05196
Alanine	A	Ala	71.03711	71.07884
Serine	S	Ser	87.03203	87.07824
Proline	P	Pro	97.05276	97.11672
Valine	V	Val	99.06841	99.13260
Threonine	T	Thr	101.04768	101.10512
Cysteine	C	Cys	103.00918	103.14484
Leucine	L	Leu	113.08406	113.15948
Isoleucine	I	Ile	113.08406	113.15948
Asparagine	N	Asn	114.04293	114.10392
Aspartic acid	D	Asp	115.02694	115.08864
Glutamine	Q	Gln	128.05858	128.13080
Lysine	K	Lys	128.09496	128.17416
Glutamic acid	E	Glu	129.04259	129.11552
Methionine	M	Met	131.04048	131.19860
Histidine	H	His	137.05891	137.14120
Phenylalanine	F	Phe	147.06841	147.17660
Arginine	R	Arg	156.10111	156.18764
Tyrosine	Y	Tyr	163.06333	163.17600
Tryptophan	W	Trp	186.07931	186.21328

Amino Acid Chart- Notice Tyrosine!

Question: Yellow Melanin or Pigment is the 1st stage of melanin how so?

Answer: Just like when you look at the flame of a fire, what color do you see 1st, the yellowish amber light. So that is the 1st spark or the initial reaction, then as the burning keeps burning the result is the fine pure **Carbon**, blackness. The Jet black, **pigment** as seen with most of our brothers and sister in **Africa** today!

As **Carbon** burns, it turns into a **black substance**, which is one of the definitions of **Melanin** "A Dark Brown or Black Animal or Plant Pigment.

Question: What is so important about Melanin?

Answer: **Melanin** controls all **mental** and **physical body activities**. **Melanin** is an extremely stable **Molecule**, and highly resistant to the digestion by most **acids** and **bases**, and is one of the hardest **Molecules** to ever be analyzed. If you do not "purify your **Melanin Molecule**", you will not heal your body of Diseases.

Question: What people have the most Melanin?

Answer: **African** people around the World have the most **Melanin** on the **Planet Earth** (**PTAH-NUN**), thus making them the "**Children of the Sun**" (**Melaninites**).

(**Dogomba Woman –
Ghana, West Africa**)

Akan Royal Family - Ghana West Africa

What is Qatum (Melanin) Physics?

Our Ancient African Melaninite Ancestors!

Question: Can your Melanin become toxic?

Answer: Yes, indeed **Melanin** can become toxic, eating the **improper foods,** having a diet that consist of a lot of **starches** and **Animal proteins**, a lot of **refined sugars,** and *basically eating food out of their natural state*, from what **Almighty Nature** has intended will have dramatic effects on your **Melanin.** When this occurs over a continual period of time, **disease (dis-ease)** manifest, also not getting the ample amount of **the Sun's (PAA RE) energy** can also have **negative** effects on your **Melanin. Melanin** is **deranged** only when it becomes **toxic. Any individual who might have toxic Melanin will act in a very similar manner, that which is Animalistic, and Barbaric. Melanin is a civilizing chemical when it is not toxic! Melanin has physical properties, and personality traits, which distinguishes it from others. That's why your body is dedicated to making Melanin.**

What is Qatum (Melanin) Physics?

Acid	Healthy Body pH Range	Alkaline
< 5.0 5.0 5.5	6.0 6.5 7.0 7.5 8.0	8.5 9.0 9.5 +

Most Acid	Acid	Lowest Acid	FOOD CATEGORY	Lowest Alkaline	Alkaline	Most Alkaline
NutraSweet, Equal, Aspartame, Sweet 'N Low	White Sugar, Brown Sugar	Processed Honey, Molasses	SWEETENERS	Raw Honey, Raw Sugar	Maple Syrup, Rice Syrup	Stevia
Blueberries, Cranberries, Prunes	Sour Cherries, Rhubarb	Plums, Processed Fruit Juices	FRUITS	Oranges, Bananas, Cherries, Pineapple, Peaches, Avocados	Dates, Figs, Melons, Grapes, Papaya, Kiwi, Berries, Apples, Pears, Raisins	Lemons, Watermelon, Limes, Grapefruit, Mangoes, Papayas
Chocolate	Potatoes (without skins), Pinto Beans, Navy Beans, Lima Beans	Cooked Spinach, Kidney Beans, String Beans	BEANS VEGETABLES LEGUMES	Carrots, Tomatoes, Fresh Corn, Mushrooms, Cabbage, Peas, Potato Skins, Olives, Soybeans, Tofu	Okra, Squash, Green Beans, Beets, Celery, Lettuce, Zucchini, Sweet Potato, Carob	Asparagus, Onions, Vegetable Juices, Parsley, Raw Spinach, Broccoli, Garlic
Peanuts, Walnuts	Pecans, Cashews	Pumpkin Seeds, Sunflower Seeds	NUTS SEEDS	Chestnuts	Almonds	
		Corn Oil	OILS	Canola Oil	Flax Seed Oil	Olive Oil
Wheat, White Flour, Pastries, Pasta	White Rice, Corn, Buckwheat, Oats, Rye	Sprouted Wheat Bread, Spelt, Brown Rice	GRAINS CEREALS	Amaranth, Millet, Wild Rice, Quinoa		
Beef, Pork, Shellfish	Turkey, Chicken, Lamb	Venison, Cold Water Fish	MEATS			
Cheese, Homogenized Milk, Ice Cream	Raw Milk	Eggs, Butter, Yogurt, Buttermilk, Cottage Cheese	EGGS DAIRY	Soy Cheese, Soy Milk, Goat Milk, Goat Cheese, Whey	Breast Milk	
Beer, Soft Drinks	Coffee	Tea	BEVERAGES	Ginger Tea	Green Tea	Herb Teas, Lemon Water

*Here is a **List of Foods** that turns your **Melanin Toxic or Purifies** your **Melanin**. The word **"Acid"** in the above chat means **"TOXIC"** and should be avoided as **Qatum (Melaninite) Beings**. The word **Alkaline** is another word for saying **Purity, Health** and **Healing**! *Refer to page# for complete list of **Acid (TOXIC)** foods and **Alkaline (PURE) foods** for your **Melanin (Qatum)**!

Questions: So, what is the Ultimate difference between Africans who have an abundance of Melanin and the European (Pale Race) who is Melanin recessive?

Answer: The Gift of **Melanin** comes from **Almighty Nature** in our ancient **TaMa-Rean** (**Egyptian**) teachings we called the **Forces of Nature** (**Neteru**). **Melanin** is a refined **chemical** located in the **body** and in the **Environment, which gives you a connection to all Nature that surrounds you. You are connected to Nature thus connected to "ALL"**.

What's unique is **Europeans** only use what is called the "**Visible**" **Light Spectrum** because "**they do not process Light or Sound within their bodies**", so they have to create *fictitious Light* from **Television** and **Radio Waves**, and try to tell you, the **African** that it is impossible to process **Sound** or **Light** within the **body**. Europeans the Pale Race had to create these things outside of themselves like **Microwaves** because they naturally do not have it, they do not have the gift of **Melanin**. *"**It is in your Melanin that the ability to process Light and Sound exist**!"*

It is the **Full Spectrum of Light** (**Whole Light**) when personifying in a Physical State manifests as **Blackness** (**Black Light**). Meaning when all forms of Lights are combined they appear **Black**. On **Planet**

Earth (**PTAH-NUN**) the solid form of light is **Melanin**. That's why Dirt the Soil of the **Earth** (**PTAH-NUN**) is **African** (**Black People**), the **Adamah** of the ground. Beings with **Melanin** in their skin are what you refer to as **Black**. However when the "**Visible**" **Spectrum of Light** appears, and manifest altogether, it is seen as **White Light**. This is no coincidence that beings on this planet called **Earth** (**PTAH-NUN**) that are **Pale skinned** are reflecting the **Visible Light Spectrum**. We know the *wavelength frequency of energy* people without **Melanin** possesses can be measured at *400 to 700* **Nanometers. That is the frequency of Light that their body is made out of.** However, you **Africans wavelength frequency Vibrates from 000.1 Miles to 3100 Miles.**

Question: What is a Nanometer?

Answer: A Nanometer is 1X (10 to the Power of -9) Meter. Wavelengths of about 450 Nanometers (**Reddish Blue**), 530 Nm (**Green**), and 570 Nm (**Greenish Yellow**) exist. Light at these wavelengths do not look pure, you cannot predict from the appearance of one such wavelength what a mixture of two or more will, should look like. The word Nanometer coming from Ancient Sumerian word "**NANA**" the **Moon God** meaning "**Sin**" and **Meter** coming from **Mother** or **Meter** in **Ancient tones**

"May-Tare", meaning **matter** or **material**, **Nanometer** is a modern term used, preferred over Angstrom used in measuring visible light, also called angstrom unit, after Anders Jonas Angstrom, a Swedish Astronomer and Physicist.

Question: What is Electromagnetism?

Answer: *According to Merriam Webster's Colligate Dictionary*:

1. **Electromagnetism** is defined as: **Magnetism** developed by a "**Current of Electricity**".

So **Electromagnetism** is **Magnetism** produced by an **Electric charge** in motion the **Physics of Electricity** and **Magnetism**. It created its own opposites. It's **Electricity in Motion** or **Energy in Motion** or **Emotional Energy**. Within the **Electromagnetic Spectrum** there are two different opposing Poles, of Energy. Both Positive and Negative; **electro (Electric)** from modern *Latin electricus* meaning "**resembling amber**" amber, being negative, and magnetic from *latin magnetis* "loadstone" being positive. **The Light** or **the illuminati** (illuminating) being **Electro** (Electric), and the **Light Spectrum** being **Magnetism**.

When we say **Positive** and **Negative** we do not mean in the sense of good or bad, we are speaking about the way the **Energy** moves or flows. **Positive** being **Outpouring of Energy** moving away from its center, and **Negative** meaning moving from the outside towards its **Center** or an **in pouring**.

Question: What is the Visible Light Spectrum?

Answer: Within the *visible Light Spectrum* there are colors, that are divided into 3 Primary Colors, and their complementary Colors which is what we see. **Yellow, Blue, Red, Orange, and Purple**. It runs from 400 Nanometers to 700 nanometers in wave length.

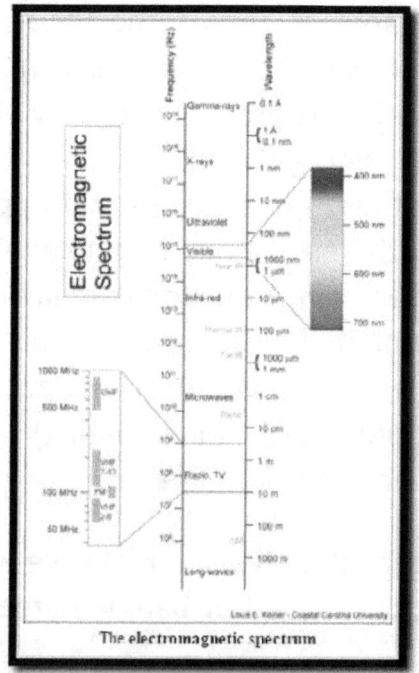

The electromagnetic spectrum

Question: Is This Light and Electromagnetic activity the same?

Answer: Yes. This **Light** and **Electromagnetic** activity is one and the same. You can determine the different types of light, by their wave length. **Cosmic rays are the shortest**, and the **longest are Electric Power**. Between these two lengths you have **Gamma Rays**, **X-Rays**, **Ultra Violet Light**, **Infrared**, **Microwaves**, **Television Waves**, **Radio Waves**, and **Electric Power**.

Question: What does all this mean?

Answer: This means **Melanin** is **Divine**, and it is a gift from The **Neteru (Nature)**. **Melanin** has **Chemical** (**AL CHEMICAL, Chemistry**) and **Physical properties**, not to mention it has **personality traits**. These traits are what makes **Melanin** Unique and different from other **Chemicals**. The **Melanin** in your **body** is like a **conductor** or **battery**, meaning it's always partially **charged**. When you're around things like **Sound**, **Light**, **Sun Light**, or **Colors**, the **Melanin** *will absorb this additional energy, and recharge itself, taking your body to a new Level!* When you are around **sounds** that are not good for your body like **Hip Hop Music**, **Heavy Metal**, or just about any kind of music today with its **artificial sounds** and **tones** it can have very damaging and draining affects on your **Melanin** and your **Body** as a whole.

Question: So you mean to tell me that our Melanin is intune with Sound, Light, and Color etc..?

Answer: Yes, as a **Qatum** (**Quantum Being**) this is exactly what we are saying! Like stated earlier, your **Melanin** can convert **Light Energy** into **Sound Energy**. That's why an Entertainer like the **Great Michael Jackson**, was such a big hit before he started to become a **Melanin Recessive Being**, then if you noticed he was not able to have any more top selling #1 albums.

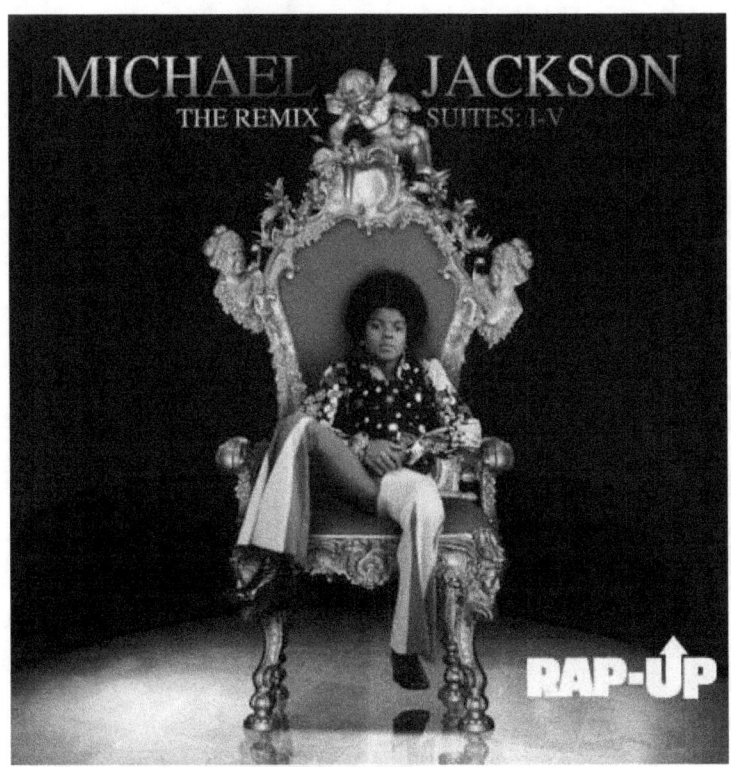

From **9 ETHER Melaninite – To 6 Ether Melanin Recessive**, the **pure Melaninite Gene Ghostize or washed out completely.** He lost his connection to **PAA NUT –The Universe (Multiverse). He SOLD HIS SOUL AND LOST HIS GIFT (MELANIN)!**

People with **Melanin** are walking Radios, and **very dark skinned people** are very sensitive to different types of **Radio frequency** or **Thought Patterns (Waves)** that are in our environment.

So, as a **Qatum (Melanin) Quantum** being everything you listen to, everything you eat, everything you do, affects you and it affects your **Melanin.**

You **Qatum (Melanin)** Beings are an **ATOM (ADAM) ATUM**, you have a **Nucleus (Solar Plexus, Central Sun)**, you have an **Electro-Magnetic Field of Energy (Electricity)** around our being called your **Halaat (AURA)**, you are **Electric** and **Magnetic** and depending on what level of **Nutational Rate** you are spinning on, that is the Kind of **Particles (People, Places, and Things)** you are "**Attracting**" (*Refer to the Law of Attraction "Nu African Mind" Vol.3*) determines your life course. **With your "Mind Power" you always have the ability to create a new you!** There are <u>Infinite possibilities</u> in the Universe, become **ALL (PAUT)** you can be, thus we say *"I am In ALL and ALL is in Me"* (ANUKI *FI PAUT WA PAUT KALUN FI NEE*).

When you tap into Higher forms of Energy through your **Emotions** which is a form of **Energy in Motion**, you tap into what people say are "**Miracles**". You have to remember, all the things **Europeans** has practiced, in order to try tap into **The Nine Stages of Etheric Forces** which the **European Race** will never be able to go pass the **6th Stage of Ether or Etheric Abode** because of the lack of **Melanin, Soul Force (BAA) the vehicle** or **Emotional Energy (Energy In Motion)** needed to raise your **Vibrations** to cause the Atoms in your body to make a **Quantum Leap** to the next **Orbital** or **Nutational Level.**

You the **Children of The SUN (PAA RE)** have the natural ability once you are in your right State of Mind (**Nine Mind, Mind in the Positive Thought Process, Reasoning**) and **Emotional Vibration**, to tap into what seems like "**MIRACLE**" Energy at will and **Mentally Polarize** your **Minds** to stay there. You must learn how to **Polarize** your **Mind** to the **Positive Frequency** and control the "**Swing of the Pendulum**", by **Positive Suggestions** which have an effect on your **Subconscious Being (Involuntary Forces – The World of your Ancestors)** which is linked to the **Plane of Force**. **The Plane** linked to the **Forces of Nature.**

Electromagnetic Energy and Melanin

As previously stated **Melanin** is the most important substance in the Human Body. It is an oxidized form of RNA, which enables the body to coordinate the production of proteins needed in cellular repair. Where ever there is cell damage **Melanin** is seen surrounding the site, functioning as a **Neuro-Transmitter** in coordination with **Melanocyte**, protein production for the repair of Damaged **DNA**. **Melanin** is a pigment in the **Skin of African People**, which is produced by **Melanocyte** Cells, and deposited, in the **epidermal tissue**. **Melanocytes** are **Neuro** like cells which produce **Melanin** and numerous proteins in response to **Electromagnetic Radiation**.

The Production of **Melanin** starts with the conversion of **Tyrosine** by the enzyme **Tyrosinase** to **5,6-Idole quinine**. **Tyrosinase** is a **Copper**-containing **enzyme** which catalyzes the conversion of **Tyrosine (an Amino Acid)** and stabilizes the conformation of the **Melanin** structure. The Metal Ion acts as a back bone for the **Polymer structure of Melanin**, resulting in a **metal** – organic complex. The Amino Acid forms **Peptide** linked formations with the metal ions. The **Ligands** are attached at the Nitrogen Atoms.

The proposed structure resembles a swastika (an ancient **African Symbol** of the "**Black Sun**" the **Etheric Energy** behind the **Sun** "**Anun-Re**").

 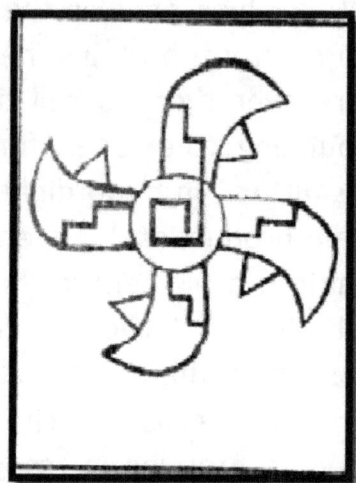

Ancient Symbols of the "Black Sun" and Cycles of Nature

The Structure has interactions occurring between the central **Cu (Copper) Ions**. This complex metal compound is the only substance in the body that qualifies as an **Organic Semi Conductor**.

Black and Brown **Melanin Granules** are oval in shape, forming a small **Dipole Antenna**. The field due to one **Dipole** can induce a **Dipole** in another **Melanin** granule nearby. **Melanin** granules act as tiny primitive eyes, forming a large **neural network structure**, whose function is to absorb and decode **Electromagnetic Waves**.

Melanin is the <u>Solid form of Light</u>, and **Melanin** gives you the ability to transduce the **Whole Light Spectrum (Electromagnetic Spectrum)**.

The meaning of the words **Dipole Antenna** is:

Dipole: *Greek* (*Diploos* – **Double**): Cancellous bony tissue between the external and Internal Layers of the Skull.

Antenna: a special **sensitivity** or **receptiveness**.

So within the **Human Skull** there is literally sensitive **"receptors"** that are receptive to **Electromagnetic Light**.

Neural – Network computers are learning machines which are made with a number of **receptors** that can adjust their weights (**Quantitative Properties**) to produce a specific output. The **Physiological** make up of **Africans** contain massive amounts of **Melanocytes** that encode all life experiences in their **Melanin** production, with the aim of <u>creating an actual reality state after death</u>. During life, visions appear frequently and **ESP (Extra Sensory Perception)** is common.

As a **semiconductor Melanin** has an energy gap. The absorption of energy is required before electrons can jump into the conduction band and make **Melanin** conductive. An increase in conductivity increases the sensitivity of **Melanin** to the **Electromagnetic** world of **Etheric Beings**, **Astral Projections**, and **Spiritual Entities**. **Melanin** is the most important substance in the **Human Body**. It is an oxidized form of **RNA**, which enables the body to coordinate the production of proteins needed in cellular repair. Where ever there is cell damage, **Melanin** is seen surrounding the site, functioning as **Neuro – Transmitter** in coordination with **Melanocyte** protein production for the repair of damaged **DNA**.

The **"Photochemical"** properties of **Melanin** make it an excellent **"Photo"** protectant. **Melanin** absorbs harmful **UV-Radiation** and transforms the Energy into harmless amounts of heat through a process called **"Ultra fast internal conversion."** This property enables **Melanin** to dissipate more than 99.9% of the absorbed **UV Radiation** as Heat and it keeps the generation of free radicals at a minimum, thus this prevents indirect **DNA** Damage.

Question: What does the word Photo mean?

Answer: According to the Merriam Webster's Collegiate Dictionary,

- **Photo means**: Coming from the *Greek* Phot, **Photo Light**, *Radiant Energy* (*Photon*)

- **Photo-Graphy**: The ability to transducer Light into an Image.

- **Photo Electric**: Involving, relating to, or utilizing any of various Electrical effects due to the interaction, of radiation (As Light) with Matter.

- **Photogenic**: Produced or Precipitated by Light, producing or generating light.

We must first "**Attune**" ourselves with the **Elements of Nature (Nazduru -Neteru)**, that are in and around us daily, through each awakening, sleeping and fleeting moment, as we elude ourselves that time is in motion. Times does not move forward nor has it come from behind, it has always been. Time does not ascend upward nor has it descended from above.

Time is now, now is the time. Time is, then now is not then. We must learn the realities of existence and reverse the process of **Self Deterioration**, both **Physically** and **Mentally**, and with that confirmed fact become a part of **ALL**. The essence of our being is that we exist and with that confirmation existence is as we are.

Time is as we are. ALL is, ALL acts, ALL does, all things are a part of ALL on into ALL. The baby of **ALL-NESS** is found in the womb of "**Quantum Physics,**" born inside – out in Etheric Existence. "<u>**Our Melanin absorbs everything around it. We are sensitive to energy**</u>". **Mental Energy** is powered by **Electromagnetism** – "**Thought waves are on different frequency**". You can have **Amber thoughts** to **Ultra Violet**, which is reflected through your **Body Atmosphere** called your **Aura**. **Whole Light Beings** vibrate threw the **Whole Light Prism** , <u>**Light produces Sound and Color**</u> and **Color** really is **Sound** , **Vibration, Tone, Octave,** and **Frequency**. **Electro Begins** vibrate threw **Electromagnetic Energy** and **Frequency Beings** threw **Octaves, Tones,** and **Sound**.

Melanin, Tyrosine and Phenylalanine

Tyrosine is a *nonessential* **amino acid,** which means that it is <u>**manufactured from other amino acids in the liver**</u>; it does not have to be obtained directly through the diet.

Tyrosine is only generated from one essential amino acid - phenylalanine.

Tyrosine is the **immediate** precursor to the **thyroid hormone thyroxin** and **Melanin**, which is the pigment in the skin that allows for tanning pigment in skin and the black pigment in hair called **Melanin**.

Deficiencies of **Phenylalanine** or **Tyrosine** may result in interruption of the synthesis of **Melanin** or **Thyroxin**.

Tyrosine is an **immediate** precursor of the <u>**Dopamine** family of hormones</u>. **Dopamines** are synthesized in the **Adrenal Medulla** and **Central Nervous System**, and <u>regulated central and peripheral nervous system activity</u>.

These **Hormones** include **Adrenaline** and **Noradrenaline**. **Noradrenaline** and **Adrenaline** are related to the proper metabolism of **Tyrosine** and **Phenylalanine**. **Tyrosine** is also metabolized to **Catechol** derivatives, which may play important roles as "**Neurotransmitters**". The route of formation of the **Catecholamines** is through change of **Tyrosine** into **Tyramine**, and the subsequent conversion to dopa (dopa=**dihydroxyphenylalanine**).**Tyrosine** which is transformed into a second precursor **Dopaquinone** via the action of **Tyrosinase**.

This compound is effective in treatment of __Parkinson's disease__. (result of the lack of *dopamine in certain regions of the brain*). The major route of degradation of **Tyrosine** is the conversion to *Parahydroxyphenylpyruvate*. This compound is then further degraded by an enzyme called **dioxygenase** to **Homogentinsic Acid**. The **Thyroid Gland** is rich in iodide. This reacts under the influence of a **Peroxidase enzyme to iodinate Tyrosine** to form the active **Thyroid Hormones**.

Tyrosine is also metabolized to **catechol derivatives,** which play important roles as **Neurotransmitters.** The major route of degradation of **tyrosine** is the conversion to **Parahydroxyphenylpyruvate**. This compound is then further degraded by an enzyme called **Dioxygenase** to **Homogentinsic Acid**.

The metabolic breakdown of adrenaline and **Noradrenaline** occur by way of an enzyme called **Monamine Oxidase** (MAO), with the ultimate excretory product being **Vanellic Acid** which is excreted in the urine. The second breakdown route for these neurotransmitters is by way of **Catecholamine O-methyl transferase** (COMT), a very active enzyme in neural tissues.

The Brain Center with the Deepest Pigmentation is the Locus Coeruleus or "Black Dot"

The **Locus Coeruleus** (**Black Dot**) supplies the **Pineal Gland** with **Norepinephrine**. The **Pineal Gland** controls the flow of **Melatonin** during the night hours to activate R.E.M sleep which allows us to communicate with **internal memory pools** or other **dimensions of life** in nature. **Melanin** also causes **Serotonin** to flow more effectively in the waking state so to "**EXPERIENCE**" more spirituality. This also helps to keep spiritual intunement at an apex. The less **Melanin** in an individual, the more calcified the **Pineal Gland** and less access the individual has to the spiritual world. **Melanin** is also used to make the **Black Dot** more in touch with the Universe. This **Black Dot** (**Third Eye**) was seen by our **Egyptian Ancestors** as the access point to **inner wisdom** and **divinity**. This was the invisible door to the pyramid which when activated would decipher the mysteries. **Melanin** absorbs light rays and stores them so that they can be used as Energy later on. This is why **Melanin** dominant people are able to use **sunlight** more effectively. A perfect example of **Melanin's** use is related to **Vitamin D**.

What is Qatum (Melanin) Physics?

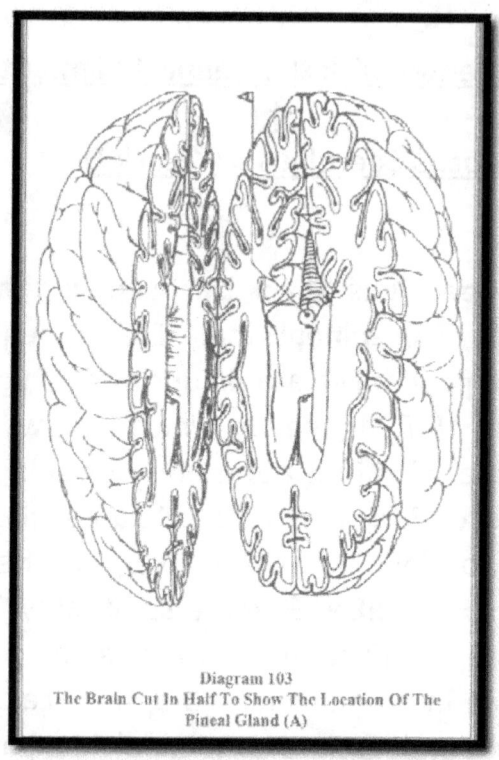

Diagram 103
The Brain Cut In Half To Show The Location Of The
Pineal Gland (A)

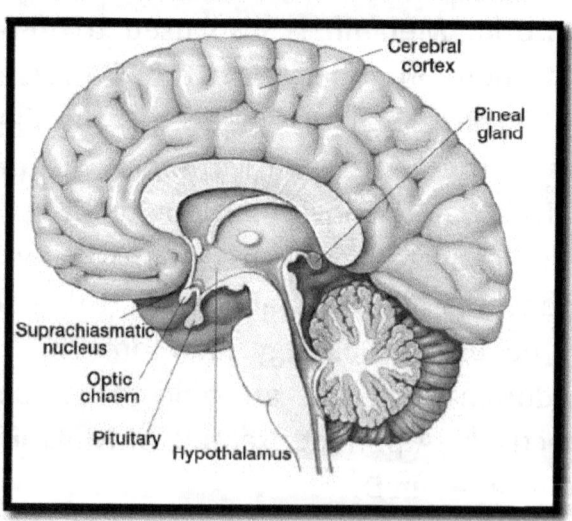

Cerebral
cortex

Pineal
gland

Suprachiasmatic
nucleus

Optic
chiasm

Pituitary Hypothalamus

PINEAL GLAND - THE BLACK DOT

Vitamin D can be found in the skin of Melanin dominant individuals after Sun exposure, whereas less "Melanated" people require the intake of *dairy products* to secrete Vitamin D. For our people, eating diary leads to blockage diseases in the body. Also another interesting bit of information about Tyrosine is it has also been suggested to be useful for the management of certain Nervous System disorders which are associated with low Dopamine output such as depression. In those depressive patients who do not respond to Phenylalanine or Tryptophan supplementation, a trial with Tyrosine might prove beneficial.

Melanin in its most concentrated form is Black. It is Black because its Chemical (AL Chemical) Structure will not allow any Energy to escape once that Energy has come in contact with it. This gives us an insight and shows that Melanin Dominant people do not require the same amount of Minerals and Nutrients in their diet as People with less Melanin. Knowing all this about Melanin is very important because you are a Qatum (Melanin) Being by Nature! We must know "who" and "what" we are and our Potential as Masters of the Universe by ordained by "*Nature*".

Question: What is the relationship between the Universe, Melanin and the Pineal Gland?

Answer: The **Pineal Gland** is the core regulator of how much of this biological life you are able to resonate with at any particular time, and it actually is on a cyclic interchange. So we know for example in our Solar System there appears to be what People have been trained to call daytime and night time, and we know there are certain things that happen during daytime and night time, because as frequency of light changes the temperature also changes. And so with the **Pineal Gland** being a regulator, of the modulation of the activity of **Melanin**, this it does, it is the modulator and it is a regulator for the activity of **Melanin**. So the **Pineal Gland** does two things primarily as far as chemical activity. Daytime it creates a chemical known as **Serotonin** and during night time (**Shadow Hours**) it creates a **Chemical** (**AL CHECMICAL**) known as **Melatonin**. **Serotonin** is needed to allow the cells to begin to become very active in discharging waste, and absorbing nutrients.

At night time **Melatonin** comes out in **darkness**, and **the Brain** has to be sealed off from light, so the eyes have to be closed, to activate **Melanin** stimulating **Hormones**. This stimulates the process of **Melatonin**; from the **Pineal Gland Melatonin** is released into the blood stream, to cause the cells to now begin to use the nutrients during the day and to rebuild the cells. So literally you're supposed to die during the daytime and be regenerated during the night.

"**Light is information**" and the Television for example, the TV wave is a band or an aspect of the **Whole Frequency of Light** called the **Electromagnetic Spectrum**. Being that you have **Melanin** you are able to pick up on **Higher Light Frequencies** coming from the **Whole Light Spectrum**, and our brains with the cycle we are now in, called the *Solar Cycle of Re(Ra)* (*refer to The Solar Cycle of RE Master Key Vol.2*)are developing into an **Electromagnetic** Computers. We have the ability to send "**Thought Forms**" in the form of pictures across space to an open receiver, and these thought forms work just like Television Waves but travel "faster" than the Speed of Light.

Therefore other Races on the Planet that do not have an abundance of **Melanin**, had to create artificial means of tapping into these higher forms of the **Whole Light Spectrum** because they could not naturally tap into these frequencies. What we need to realize with our Real Eyes (3rd EYE) is that a picture is a frequency of **Light Wave**.

In cultures where people are **Melanin** dominate they had these pictures, as relics, as **Hieroglyphs**, and as Codes, that signify that our Ancient Ancestors where always pulling from a *"higher spectrum"* from **Electromagnetic Frequencies**. We have the ability, to see a picture in our minds and send a thought to a person and when a person is open to receive the thought wave, they will be able to see this image in their minds.

In other words being we as Ancient **TaMa-Rean (Egyptian) Melaninite-African People** we have the ability to transfer **thoughts** in the form of pictures to another person without the aid of any type of exterior machine or device. Now the important things to remember is if you do not use your gifts, your gifts to visualize, to hear with your inner ear, or inner feelings (inner awareness, inner sensing), you will lose these gifts or these gifts will go dormant.

You have to remember everything is **Light**, and what we are speaking about is **Atoms (Atum)**, known as **Photons**, and the **Vibratory Rate**, at which these **Photons** are *pulsating*, is determining the colors we see, or determining the attributes of that particular frequency, that **"Photon Vibration"** creates.

The **Pineal Gland** is the **Modulator** that is able to pick up on all the frequencies that are around us. **Sound** becomes **Light** and things manifest, incarnate, materialize. As sound of music makes you feel, makes you move, makes you wish to sing or respond or resound, resonant, re-spawn or respond to vibrations of sound. We as **African People** who have a supreme abundance of **Melanin** can grasp the **Whole Light Spectrum**.

We must remember that everything is **Light**; everything is vibrating at a particular rate, what appears to be solid is really not solid, and depending on if you can slow your atoms down to that vibratory rate depends on if you can merge with it and walk through it. So what we are speaking about is Modulations and **awareness of light, light frequencies and vibrations.** When you see **Full Spectrum Light** in its physical state, it manifests as blackness. That is, all light and color combined are blackness. And the manifestation of the solid form of light is **Melanin**.

Melanin gives you the end road to be able to interact and to know the different frequencies, that light can manifest itself upon. And when we speak about light know we use the word "Light" for lack of a better word. **Light really is different moods of Vibrational Energy.** That means minute **Particles**, that are coming from the Sun at us at different speeds and you call them **Light**.

Melanin - Sun Heat Genes

Sun Heat Genes exist within the **Bone Marrow**, and your scientist or medical personnel today refer to it as the **Mast Cells**. These **Mast cells** are responsible for utilizing certain elements, like potassium, calcium, boron certain elements for you to grow healthy. Our brothers and sisters today in our Home land Mother and Father land **Africa**, you can still find very dark brown to black to blue black **Africans**. This being our brothers and sisters **Genes** are so potent, because of these **Sun Heat Genes**. As **9ether Beings** "**Sun Heat Genes**" causes you to burn outwardly, and it makes your hair curl up or what they call kink up, or king up, because when you take a sample of your hair and hold it next to a flame of fire your hair will curl due to the **9ether Gases** within your hair. So you as **9ether Beings** are so potent because of the amount of **Melanocytes** you have in your skin and **Sun Heat Genes** (Sun Codes) known as **Melanin**. **Melanin** is **Sun Heat Genes** and the solid form of **Light is Melanin**. Your **DNA** is coded in **Melanin** & your **Supreme Melanin (Brain-Neuromelanin)** has to be activated in this day and time. **DNA** is the Soul (BAA) "which when triggered opens in awareness."

Soul **(BAA)** Seed **(KHAT –BODY)** you have to reach down in your **DNA** and pull divinity out. Your **Spirit (KAA)** is in the **RNA**. Your **African** self **(Body)** as you can see is a small Universe within a larger Universe thus **The All in All**. Our Bodies are a tiny Network of **Electromagnetic Energies** radiating at different Speeds and **Moods of Vibration** within a Larger Network – **The ALL (PAA PAUT)**. We must begin to realize that we are deities and through our DNA (Genes-Genetics), **we have no reason to be afraid, reach down in your genes and pull out divinity!** Through the **Power of Thought, (in Egypt the Deity - Neter of Thought was known as Thoth or Tehuti.) Mind Power. "The Journey starts with a Mire Thought!"** Your **Potential** is opening and grasping the **magnificent code** (**DNA**) that we all have been endowed with. When you are not elevating with time then you are dead.

Question: What is DNA, and why is it important?

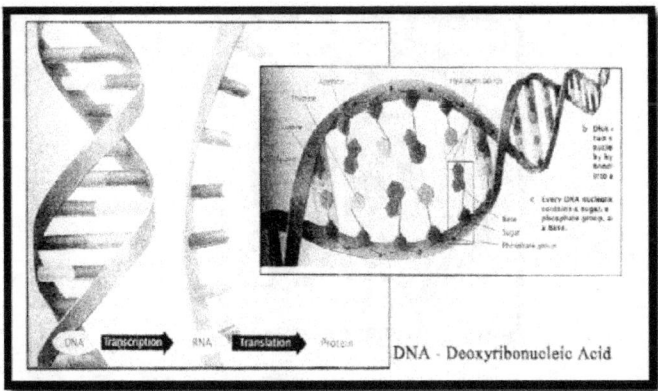

DNA - Deoxyribonucleic Acid

DNA – Deoxyribonucleic Acid

DNA is so important because within your **DNA** through what is called "**Heredity**" there is a recorded past of everything that has ever happened to a person and there **family Lineage** (**bloodline**) throughout time, leading to the first groups of people on Earth. As we know, **all life started in Africa!**

With **DNA** testing many are able to find a direct **blood link to all Ancient** and **Great Rulers**. Many well trained **Master Minds** throughout history have been able to tap into infinite resources dormant in their **DNA**, so "**NOW**" is your time to tap into your **Hidden Potential** and true **Powers of your Mind**, so that you can once again take your rightful place in the world. In order to know where you going you must know where you came from, **Sankofa**!

You're Electro-Magnetic Body!

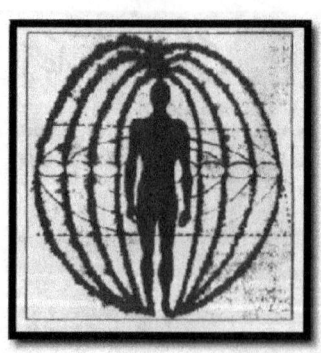

Question: Do Human Beings have an Electromagnetic Field?

Answer: Yes! There are two currents of **Electrical Energy** that run through your body creating a "**Magnetic Field**" called your **Haalat** or **Aura**. Your **Electro-Magnetic Field** (Aura) is a part of Earth's (**PTAH-NUN**) **Magnetic Field** in which your **Aura** (**Electro-Magnetic Field**) moves independently. These fields are composed of **Negative** and **Positive** Vibrating Energies or Forces. These forces are known as:

Centripetal Force – That comes towards the Planet Earth (PTAH-NUN) from the universe (i.e. Energy from the Sun, Moon, and stars or other Solar Bodies.)

Centrifugal Force – Force that comes from (within) the Planet Earth (PTAH-NUN), (I.e. energy from plants, minerals, the soil, and water).

Human Beings Aura (**Halaat**) contains your Spirit (**KAA**), Soul (**BAA**), and Physical (**KHAT**) form. The Halaat (**Aura**) which is the "Life Force" (**Ankh, Sekhemic Energy**) of the Body may be looked upon as the *Atmosphere* of the Body, because your **Aura** reflects the "Real" vital Body, Mental Mood, Emotional Energy, Health and Character.

Notice the word **"Atmosphere"**, which is very key to knowing your relationship with Planet Earth (**PTAH-NUN**) and ALL (**PAUT**).

The word *Atmosphere* is defined according to the Merriam Webster's Colligate Dictionary as: (N) Greek Atmos – "Vapor", Sphaera, "Sphere". (*The gaseous envelope of a celestial body, "as a planet"*), a surrounding **influence** or environment.

The word *Influence* is defined as: (Influere – *to flow in*) from Fluere "*to Flow*".

- An *Ethereal (Etheric) fluid* held to flow from the stars (*which are Suns*) and to affect the actions of Humans.
- An **Emanation** of occult (hidden) power held to derive from stars (SUNS).
- An **Emanation** of Spiritual or moral force.
- The act or power of producing an effect without apparent exertion of force or direct exercise of command.
- The Power or capacity of causing an effect in *indirect* or *intangible* ways: To sway

The word *Emanate* is defined as: from "Manare" **to flow**, to come out from a **source**. *"Emit"*.

So according to these definitions your **Aura** which is the **atmosphere of the body,** can be seen as having an *"influence on your environment"* via your **"thoughts"** which is your mental state and mood which is your **Emotional Energy** or **Energy in Motion**, which is **"Electro-Magnetic"** in nature. So at times depending on your Emotional Moods you can have a **Aura** or **Atmosphere** around you of **Positive (9ether)**, life giving forces (energies), or **Negative**, draining **(6ether)** death causing gases (energies). So these different layers or Moods of Vibration (*Emotional Energy*) are reflected in colors, which may appear cloud like or bright and clear according to the **Emotional Mood** a person is vibrating on. The colors and moods of vibration fluctuate depending on a person's mental state, and mental attitude. **Positive Mental Attitude (Thoughts)** reflects **bright, clear** and **vivid colors** which attract **9ether** life giving, vitality particles.

A Person's **Aura** can be felt, which is why you have heard the statement many times **"that person gives off a good vibe,"** The word **Vibe** equals **"Vibration"**. Some people are a pleasure to be around, they seem to revitalize, and invigorate everyone they come in contact with. These are your life giving people who have Positive Etheric particles within their Energy Fields or **Aura**. All they attract is good and Positive. These are the people who always seem lucky or always seem to be upbeat and have a Positive outlook on everything. They are upbeat and full of life!

Then you have the person who seems to *"drain your energy"* when you come in contact with them, you have stepped into a **Negative Aura Atmosphere** (Energy field), an **Aura Atmosphere** full of **fear**, **doubt**, **worry**, and **anxiety**. After being around such energies emitting from these kinds of people, you start to doubt yourself, and your confidence is undermined. This means there **Aura** is spilling into yours, and they are sucking your energy, such people in this day and time are known as "**Energy Vampires**". They seem to complain about life and all the Negative things that happen. They love to play the victim role or "**why me**", "**poor me**" role. These kinds of people have crystallized a **Negative Aura Field** around themselves **Subconsciously**, which remember, is reflected by your **Mental Attitude** (**Thoughts**). Having **Negative Thoughts** attract these Negative 6ether chaotic Energy Forces within your **Auric Fields**.

These types of people if you are not able to raise their Mental State, thus raising their Vibration or at least teach them how to stay on the "Positive Life," frequency should be avoided at all cost.

In Ancient times **Etheric Forces** were seen as *"Little Angelic helpers"*, or *"Demonic Hinders"*, all **Attracted** to us by the **Thoughts** we are keeping and charged by us through our **Emotions**. These **Etheric Forces** are little microscopic cosmic forces vibrating on a much faster vibration thus not detectable by the naked eye but can be seen living and animating and taking form through People, Place, Things, and events or situations.

Question: As Qatum (Melaninite) Beings are we linked to ALL that is around us?

Answer: Yes, as **Qatum** (*Melaninite Children*) Beings you are linked and influenced by your environment, you are linked to the Plants, for Plants have an **Aura** or **Energy Field**, Animals have an **Aura**, Trees, Bugs, Insects, Dirt, Water, all emit an **Aura** and a Mood, a Vibration. Again what we will start to realize in the coming years leading up to 2012 and beyond, you are Qatum (Melaninite) beings, your **Aura** is "**infinite**" and "**boundless**", your **Aura** is directed by the **Powers of your Mind**. Everything has an effect on everything. No matter where you are on the Planet

Earth (**PTAH-NUN**), in the Universe or Galaxy. Planets have an effect on other Planets; Solar bodies have an effect on other Solar bodies, thus we are **ALL within ALL**.

Light – the Birth of Emotion (Energy in Motion)

*"The **Light** of the Body is the **Eye**: if therefore thine eye be single' thy whole Body shall be **full of Light**." –* Mathew 6:22

Question: What is Light?

Answer: **The Sun** (**PAA RE**) and other forms of Light give off vibrating Energy that our **Eyes** can see called **Light**. Light is defined by Wavelengths, which is measured by the distance from one peak, or top of a wave of Light of Energy, to another. We see different speeds of **Vibration** as different Colors. When the **Energy Vibrates** really slow or fast we can't see it at all. Very slow **Vibration** makes Radio Waves. **Microwaves** and **X-Rays** are very fast **Vibration**. You have **Radio, Red, Violet** and **X-Ray**, which range from 4 to 380 Nanometers. Microwaves, a very High Frequency Range 750 Nanometers. A Nanometer is a modern term used, which is preferred over Angstrom used in Measuring **Visible Light**. When all different **Colors of Light** called "**Prisms**" from the Latin word "**Prisma**" are mixed together our Eyes see this as

"**White Light**". When you see Full (Whole) Spectrum Light in its Physical State. It manifests as **Blackness**. That is, all Light (Whole Light) and Color combined are **Blackness**, and the manifestation of the Solid form of Light again is **Melanin**. When While Light hits any Colored thing, some of the Light is absorbed and some of the Light is bounced off or "reflected" toward our Eyes, and those reflected Colors are what you see as different Colors. So you see its all an "**Illusion**". Light is not Light but **Energy** moving in Darkness as Vibrating Waves at different speeds.

Thus the decision to decide what is ugly, is a Personal choice; to see **Colors** and to mistakenly prefer one over the other, or this Person over that Person, Like and Dislike, caring and Uncaring, Concerned and Non Concerned, Lover and Hate, War and Peace, all become manifest through the Light of Knowing, Knowledge. When People see they want to see more. Such as, in Inner Sight (Spiritual, Inner Awareness) one gets the assurance whether their Eyes are opened or closed of existence through feeling. Thus, the **True Principle of Neter** (**Nature, Deity**) is not **Seen**, but "**Felt**".

Once one opens their Eyes, they get the **illusion of Reality**, thus they need imagery to send messages of things they know or know of, to determine Neter

(**Deity-Nature**). So the "concept" that Light is good and Darkness is bad, is one of the "ultimate deceptions". The different Monotheistic Schools of Thought mainly **Judaism**, **Chirstism**, and **Muhammadism** teach you to stay in the Light. They tell you that Darkness is ignorance. Yet the GOD that created the Light was in Darkness before he created the Light. Now think, what is more "**Omnipresent**" the Light that is confined within a space within a sphere or the Darkness that is "**Infinite**" in the Boundless Universe? The Darkness in the Universe of course! You can see Light in Darkness but you cannot see Darkness in Light, thus Light is an "Illusion".

Question: Is there a connection between the State Blackness and PAUT (The ALL)?

Answer: Ok you are speaking about the State of Blackness and **PAUT** (**ALL**), right? Ok what all Esoteric Schools of Thought are calling **PAUT** (**ALL**) precedes **Light**, because **Light** is a "thing", and let me explain what we mean by this. There is no such thing as what we have been taught is "Light" or "the Light". When we were taught about "**Light**" what they are teaching us is different Moods of Vibrational Energies, that means minute particles that are coming at you from the **Sun** (**PAA RE**) at different speeds, and you call these minute particles Light.

There is no such thing as the **Light**, when you start to view this concept of light from a **Higher Mind (Awareness)**, **Scientific Mind** and **Quantum Mind (Nine Mind)**.Then you will begin to realize there is Energy coming from the **Sun (PAA RE)** and we are picking it up and have been taught to call it **"LIGHT"** for lack of a better word. It's not a **Beam of Solid Mass** but <u>**Particles of Energy**</u>, so therefore what we are calling **Light** is not a Mass, a Whole. **Light** within itself breaks down into **particles**, so **The ALL (PAA PAUT)** cannot "be" **PARTICLES**, because it cannot be part, and **Particles** are portions. **The ALL (PAA PAUT)** is, now darkness is, and darkness can't be **Particles**, because **Light's Particles** manifest in darkness's existence as **the ALL (PAA PAUT)**.

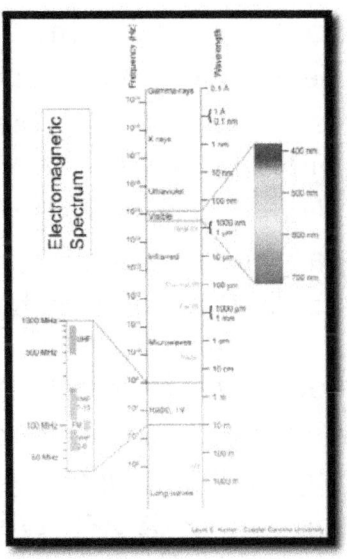

So to answer your question, yes **PAA PAUT (THE ALL)** and **Darkness is ALL**, your **GOD is Light**, and your **GOD** or **ALLAH**, or **Jehovah (YAHWEH)** is a **"Particle"** in the **Darkness**. Because **GOD, ALLAH, YAHWEH**, is talking from the **Darkness** about what he is going to do in the **Light**. SO you're

GOD, ALLAH, and **YAHWEH** is at a specific point in **Space** and **Time** having an effect on **Matter**, when he says "**let there be light**" or "**let Light exist**" and your **GOD** makes reference in the Bible about the light as the "**SUN**" in ancient Egipt we say **PAA RE (The Sun)** or **Amun-RE** or **Amun-Ra.**

So now the **Sun (PAA RE)** is **93,000,000.00 miles** from **Earth (PTAH –NUN)** varying and **the Sun (PAA RE)** is "there" and it's sending these **rays of Particles** toward us, the **Light** that **GOD, ALLAH, YAHWEH,** said let exist He or She let exist at a specific point in **Space** and **Time**, and on the 3rd planet from **the Sun (PAA RE)** we are **93,000,000.00 miles** away from that **Light** that He or She said let exist. So your **GOD, ALLAH,** or **YAHWEH** designated a place for **Light**. So **Light** just doesn't exist it's there, when you begin to start to view persons, place and things, on a **Quantum level** as a **Qatum Being (Melaninite)** and our connection with **ALL (PAUT)**, then you being to step pass the blinding light of Religion and Religious concepts that has enslaved and trapped the minds of **Qatum Beings (Melaninites)** for too long.

You have to reactivate **Brain Melanin, Supreme Melanin** called **Neuromelanin** so you will be able to receive whole light particles of energy coming from **the Sun (PAA RE)**. **Whole Light Particles** of Energy packets traveling towards our planet on a constant basis have within them what we term as "**Light Messages**". These **Particles** which we have been taught to call "**Light**" carry messages, which is why the word "**Angles**" and "**ANGELS**" look and sound phonetically similar. The Angels, religion was speaking of was really different **vibratory rates** of the **Whole Light Spectrum**, different frequencies.

In most Esoteric schools of thought **Melchizedek** also known as the **Angel Michael** sacred colors are known as **Violet (Ultra Violet – UV Ray = Black Light)** which is said to manifest on this side of the **Visible Light Spectrum** as **Green Light**, seen in plant life symbolizing "**HEALING**". **Malachi CH 4**, and the Hebrew word **Malachi** means "**MY ANGEL**" and in **Malachi chapter 4:2** "*But unto you that fear my name shall **the Sun (PAA RE)** of righteousness (**MA'AT**) shall arise with Healing (**Green Light Energy**) in his wings*".

Question: What Light Frequency is the Soul (BAA)?

Answer: The "**Light of the Soul**" goes back to what the **Ancient Tama-Rean (Egyptian)** called **Nu** or **NUN** that Primeval (First) Energy known today as **Black Light (UV-Ray or Ultra Violet Light)**. Your **Soul's Light Frequency** is from the **Original Creative Forces** called **NuPu – NUN (Nine Ether)** that birthed all other

Energies and Gases throughout the boundless Universes. Remember **YOU** are on a journey to that **"Source of Energy"** in you, that **Black Light** or **Black Dot**.

Dark Light (**Black Light**) travels faster than the **Visible Light Spectrum**, faster than the **Speed of Light**. The Speed of Light differs in each solar system which is determined by that Solar Systems **"Solar Body"** (**Sun**) but **"Thought travels faster than Visible Light**," and **Emotions** or **Energy in Motion** or **Amber Light** travels faster than **Green Light** (Dark) **Intellectual Light**. Intellectual light is Healing Light. It travels slower because you have to sit back and think before you act on anything which slows down the flow of Energy, but we tend to flare up quicker. Learn to channel your **Amber Light (Negative, Lower Desires)** to the **Positive Frequency** then you are **GOD** (Deity) in control of yourselves.

Black Light (NUN) = Pure Facts
Green Light =Health, Holistic Healing, Nature
Red (Amber) Light =Pleasure, Desire, Enjoyment

Question: If Light is an Illusion then what is Sight?

Answer: According to the Merriam Webster's Dictionary **Sight** is defined as: *the **Physical Sense** by which **Light stimuli** received by the **Eye** are **interpreted** by the **Brain** and constructed into a representation of the Position, Shape, Brightness, and Color of objects in Space. **Mental or Spiritual Perception.***

Sight is one of the five external Physical Senses that Humans have. Contrary to what we have been taught the "Eyes" do not actually see, the Eyes contain receptor cells which respond to messages from the Brain as they experience external stimuli. They work by **touch**, the "**one real sense**". Receptor cells are sensitive to specific classes of stimuli within a certain range of intensity or vibration. To explain stimuli (plural for stimulus) – Is external interference or action upon receptor cells. **Light**, for example, is a stimulus for **Vision**. Your ability to see depends upon a certain amount of **Light** penetrating the lens and cornea of the Eyes. So that it focuses on the delicate retina. The part of the Eye has nerve cells which relay impulses to the **Cerebral Cortex** of the **Brain**. Thus an Image is **Visualized**.

Question: So how do the "Eyes" function?

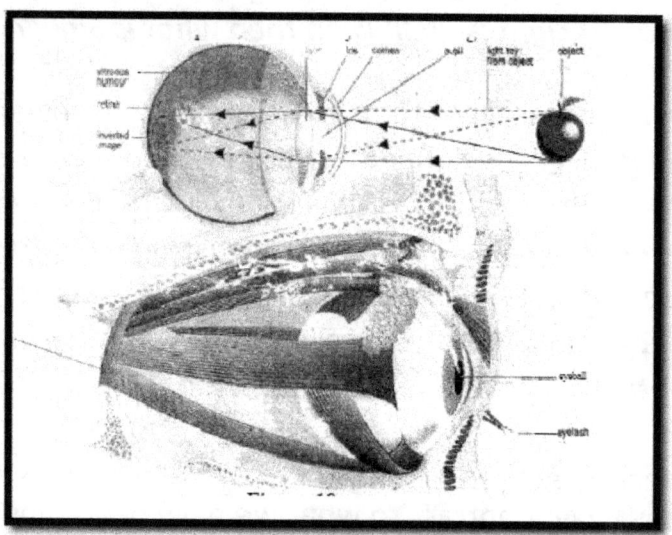

Answer: Your **"Eyes"** are often said to be the windows of the **"Soul"**. The Eyes are spherical structures anchored by six muscles into bony sockets in the Skull. Under the direction of the Brain, these muscles cause the Eye to track moving objects. The inner mechanism of the Eye has a superficial resemblance to that of a Camera; Light Rays are focused through a lens on- to a light-sensitive Area, **The Retina**. But this camera is a highly flexible one, and it's attached to a *Supercomputer* (**The Brain**). The **Retina** itself is made of about 130 million receptors, and the Eye can distinguish **8 million colors**.

The "**EYE**" is moved by six voluntary muscles and is kept firm by transparent fluids within it. The Eyeball consists of three layers of tissue. The outer layer is what is known as the White of the Eye, the Black middle coat contains the blood supply and the innermost layer is composed mostly of nervous tissue. At the front of the Eye, these coats change. The outer coats become the transparent Cornea. A round opening in the middle coat is the Pupil, and behind it is the Lens. The muscular Iris surrounding the Pupil changes the size of the opening and gives the Eye its color. The Lens brings together the Light from the objects one looks at and forms a visual mental image. When you look at the word "**imagination**" you see the word "**Image**" (image-nation) in this word. So **imagination** is the creative ability of the **Mind** to form visual mental pictures or

75

images. Another word for this process of seeing mental images is called *"Visualization"*.

Question: So What is Imagination?

Answer: According to the Merriam Webster's Dictionary Imagination is defined as: *"The act or power of forming a **mental image** of **something not present** to the **Senses** or never before wholly perceived in **Reality**, Creative Ability."* So as you can see **Imagination** is the ability to create an **idea** or **mental picture** in **your mind**.

The Holographic Nature of the Brain

Question: I have heard many times the Human Brain is like a Holographic Super Computer how so?

Answer: Yes, this is true when examining the functions of the brain and the way Brain reacts to outer stimuli via senses, and the way the brain stores information, our Brain is a *Holographic Super Computer*. Just like a computer our Brains have a certain amount of Memory or Bytes that can be used as impulses from the *Senses* that are beamed to the Brain, connecting, converging and interfering as they overlap within your Brain Cells. **Electro-Magnetic Energy** serves as the *Holographic Brain's* laser-like Light. Your Eyes serve as the Object Beam. Your remaining senses and emotions serves as the Brain's reference Beams.

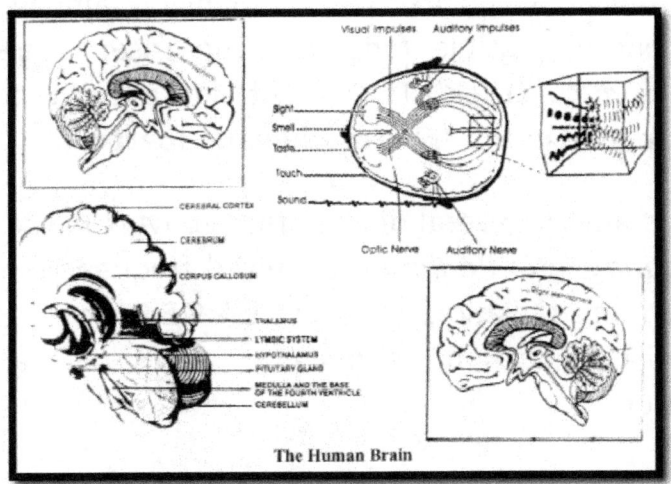

The Human Brain

Now the Science World has discovered the Technology where they can put tiny *Microchips* into the Brain and even download programs right into the Brain, (*refer to Matrix Movies Pts. 1-3*) touch screens as well as **Mind control** of Computers.

Now let's take a look at the word *"**Hologram**"*, you see *Holo* or *Hol* meaning *Whole Entire*, and **Gram** meaning *"**something written**"*, thus Hologram is a *whole* or *entire message*.

Holograms are produced by illuminating the scene with Light from a device called a **Laser**, which works off the "Principle" of vaporizing the hardest and most heat resistant materials through multiplied stimulation of Atoms. The Laser is a number of several devices that turn striking **Electromagnetic Radiation** of mixed frequencies of highly amplified and coherent visible radiation. The Laser is also called *"Optical Laser"*! The Output of a Laser is in the visible region of the *"Electromagnetic Spectrum"*.

Laser is any of several devices that convert incident *Electromagnetic Radiation* of mixed frequencies to one or more radiation. The word **L.A.S.E.R** stands for **L**ight, **A**mplification, **S**timulated, **E**mission, and **R**adiation.

Laser provides the long sought after model of how **Visual** and **Sensory** information is received stored and recalled by **the Brain**. Our **Ancient TaMa-Rean (Egyptian) Ancestors** knew all this about **the Brain**, and being we are the **"Children"** of our **Ancient Ancestors TaMa-Rean (Egyptian)** it is due time we know and begin to reawakening the dormant powers of our **Quantum Minds**.

Sound, Frequency and Vibration

Question: What is Sound?

Answer: Each time you hear a **Sound, Energy** is reaching your **Ears**. The **Sound Waves** from a **Vibrating** object make your **eardrums vibrate**. The

Vibration of the Eardrums affects nerves which lead to the Brain and you hear the sounds produced by vibrating objects.

Sound Energy can be used to cut solid objects. Doctors have used powerful Sound Waves to perform operations that otherwise could not have been done. Different **Vibrations** make you hear different **Sounds**. A faster **Vibration** makes a **Higher Sound**. **Sound Vibrations** cause **Sound Waves** in the air and these **Sound Waves** can travel through solids, liquids, and gases, all three stages or lower forms of **Matter**.

When you speak, your vocal cords **"vibrate"**, somewhat like the prongs of a tuning fork. The **vibrations** of your vocal cords form sound waves in the air that can and will affect all Matter. For instances, in our everyday life we are becoming more and more aware of the impact of spoken words, their power and effect they have on our environment and whole being. Words of love, for example bring Joy to ones being, as for words of anger can incite anger or cause pain in others and soft words spoken in times of confusion can bring ease. For these scientific reasons our **Ancient Egiptian (TaMa-Rean) Ancestors** knew the importance of **words**, **tones**, **sounds**, and **vibrations**.

In ancient times we knew that just a person's name was very important and spoke a lot about the person carrying a name. A name is really a tone which sets off a vibration in ones being that cause a chemical and emotional reaction. In **Ancient TaMa-Re**, **Egipt**, we placed a great deal of importance on the knowledge of names. When one knew the name of a **Neter (Neteru)** "**God**" or **Tuta** "**Devil**", and the soul or spirit addressed the Neter "**God**" or **Tuta** "**Devil**" by name they would be forced to answer the caller and do whatever was desired.

Also this same principle of Names and Sound can be used to stimulate and arouse the inner deity in oneself to wake up and become activated, or the activating force in one's life, which is the inner meaning of the **99 Names** of **ALLAH** or **God** in Islam, which was originally taken from the 99 Names of the **Neteru (Nature)** in **Ancient TaMa-Re (Egypt)**. These names when **Chanted** or said over and over again cause a **vibration** within oneself and arouses invisible sound currents in one's **individual Aura (Electromagnetic Field)** that can overcome the effects of **Negative Karma** that may be flowing through you.

All of this is to say as **Qatum Beings (Melaninite Children, Children of Nature)** we have the ability to process **Sound** and **Light** with our own **Body**. We have always had an **effect on our environment** and **we are connected** to all things around us. The words you speak, the company you keep and the things you see all play a major role on your **Subconscious Minds**, which is very important in this day and time.

As we are going through this **shift, in consciousness, shift in awareness** and **shift in DNA**, plus a shift as a whole Race and Planet (**PTAH-NUN**) you will begin to see that the words you speak and the vibration behind these words which produces a certain **sound** or **tone,** which have an effect on the **vibration** in the **Atmosphere (Ionosphere)** will start to manifest in your life much quicker. We are truly the shapers and controllers of our own destinies.

The more we start **thinking** on these **Higher Vibrational Perception Levels** the more we become "consciously" **aware of the presence of Higher Dimensions**. We are now at the time where our Planet (**PTAH-NUN**) as a whole is moving from a **3rd dimensional reality** to a **forth dimensional reality**. Human Beings and most importantly **Qatum (Melaninite People, Children of the Sun)** will begin to start thinking and reasoning on a more **Quantum**

level (holistic), learn to examine life, our environment and the Universe as a whole as not a separate entity but as living organism within **ALL**. We are in the day and time where we must realize that we are **Quantum Beings**, the Microcosm of the Macrocosm.

Question: You mentioned chanting Earlier, How will Chanting help the Spiritual Self or Body?

Answer: Excellent Question! Again the entire Universe is based on **Tones** and **Vibration** and certain combination of words create **tones** that can penetrate the Density that surrounds us much like when a soprano singer raises their voice to a certain key and holds it to the point that the person can literally shatter glass. This informs you that **tones** penetrate even dense Matter. They also create Heat and Energy.

So what you are experiencing is that Voice transforming from Sound into a form Energy attempting to penetrate the Glass. So Chanting can be used Negatively or Positively. When learning to speak your ancient **TaMa-Rean (Egyptian)** Language called in this day and time **Nuwaupic**, you will see that the tones of this ancient language is in synch with the positive **Nine Etheric Energy Forces of Nature**, called **NUPU(NUN)**.

Thus, the Chants you will be taught in **Nuwaupic** will also consist of **Positive tones**. Which have been selected and arranged in a way to penetrate the dense **Matter** of the **Physical Plane** onto the **Plane of Force** upward and onto the **Spiritual Plane** then the **Mental Plane** that you may be heard by your **Etheric Parents**, which are **Higher Positive Energy Forces** (within the **Electromagnetic Spectrum** of Light) that are the **Original Creative Forces** that grew all the **Universes**, thus putting you **The Melaninite Children (PAA QATUM) back intune with Nature**. Your Whole Body will become a telephone where you will start to receive more **Higher** and **Positive Mental Transmissions**, producing more **Positive** and **Successful Thoughts and Inspirations** that are for your **positive** growth in **Nature** and not your demise.

Question: You mentioned our Minds have the ability to pick up higher transmissions coming to us from The ALL (PAA PAUT), is this what happens during Meditation?

Answer: Well let's first look at what Meditation is.

- According to The **American Heritage Dictionary** for **Meditate** is defined as: *to engage in contemplation, especially of a spiritual or devotional nature.*

Meditation is simply shutting up and thinking in attempts to cleanse The **Subconscious Mind** of worthless thoughts and negative memories that interfere with the Inner Voice there to guide you. In you, **Melaninite Children (PAA QATUM)** there is Agreeable and Disagreeable Genes from both sides of your **Physical** as well as **Spiritual (Ancestral) Family**. Each person has a **30 distinct Personalities** speaking with them if not more depending on who you have come intimately in contact with. These **30 Personalities** come from **15 genetic** Ancestors that were passed on in your Gene pool from your **Mother's side** and **15 from your Father's Genes** all at the time of **Conception (Birth)**. All **30 Personalities** or **Ancestors** are all trying to have a place, and want to live through you. Some are there to aid and guide you and others are there for their own selfish means, because they may still be trapped on this **Physical Plane** of **Existence** as disembodied Souls and need your energy, to subside off of, in order to burn out certain desires so they can **Elevate** or **Revolve** to **Higher Vibrational Levels**. So the **Art Form of Meditation** helps to guide one's self back in **Attunement** and **Alignment** with the "**Real You**", those **Agreeable Forces** that are here to aid you and help pull you back up to the **Living Forces in Nature**. This is the day and time we are living in, were we must re-attune (**RE-ATUN**) our **Inner Beings** with the **Positive Forces** of **Nature (NUPU, NUN)**.

So again the **Art form of Meditation** or simply **Concentration** causes the fetters of the Mind to dissipate and alignment with more higher, agreeable transmissions from Positive forces within nature will take place.

Question: What does all this have to do with Qatum (Quantum) Physics?

Answer: Everything, because once you learn to sit still and **Meditate** or go back inside, you will begin to perceive and shift your perception from thinking as a separate being, to that "you are connected to ALL that is around you". You will begin to realize everything on this side of **H1** (**Hydrogen**) leading you back to the *Etheric realm* is all connected, thus ALL in ALL. As we are taught in **The Sacred Wisdom of the Grand Hierophant: Tehuti "Thoth"**:

- *"While **The ALL is in ALL**, it is equally true that **ALL is in ALL**. To him who truly overstands this truth, Right Wisdom has come in the form of Right Knowledge".*

Being Connected to **ALL** gives one the realization that they do have an effect on their environment, and what they think, do and act upon does shape and mold not only their lives but all those around them for their good or bad. It also gives you the Power to know that all your Divine Potential is already within you, waiting to be called upon and released in a Positive way for you to live your life's dream and purpose in order to help the World a better place.

The **Solar Cycle of Re (The Sun Cycle)** the cycle within Nature is the Cycle were all your latent abilities come to life, and were anything is now possible, where the veil between **Heaven** and **Earth** or simply **Mind** and **Matter** is becoming very thin. Your thoughts, **Mind Power** and your **Physical Bodies** as **Qatum Beings** gives you access to infinite vast resources in and throughout the **Boundless Universes**. Knowing and realizing these simple realities alone will start to do wonders in your everyday life. Just by meditating on these words alone will start to produce new type of thoughts you never thought, thus producing **Positive Inspirations** that will help guide you to your Higher Purpose and Potential during this life time or Incarnation. We must remember each Soul in this day and time has incarnated for a reason, has decided to come into physical form during these last days and times for some and a new day and time for others. We are not just our 5 senses of Hear, Touch, Smell, Feel and See but there is more to us. Stop and Think…

The Four Aspects of Sound
"*In* Esoteric Sciences" (NUN)

1. **Sound** which emerges from **The Mind**. The isolate Consciousness.
2. **Sound** which emerges from **Ether (Sekhemic Energy)** the Primal Voice of the **Black Light (Hidden Light of the Universe)**.
3. **Sound** which emerges from expression and communication.
   The Soul (BAA-Solar Plexus)
4. **Sound** which has its seat in the **Heart** area, this is different in approach to the **Chakras (Arushaat)** wherein it is related to another aspect, in sound it is the unspoken, the heart swelling with passion.

Question: What if Frequency?

Answer: According to the *Merriam Webster's Collegiate Dictionary* the word **Frequency** means: the number of complete oscillations per second of Energy (as **Sound** or **Electromagnetic Radiation**) <u>in the form of waves.</u>

Please note: that **Qatum (Melaninites) Beings** and **Human Beings** as a whole are constantly **Vibrating** on different **frequencies** within the **Whole Spectrum of Light** called the **Electromagnetic Spectrum**. Again this is reflected and can be seen through the **Electromagnetic Atmosphere** of the **Human Body** called the **Aura**. People fluctuate between **High Frequencies** and **Low Frequencies** manifesting through **Colors** seen within a Persons **Aura**. These different **frequencies** give off what is termed **"Moods of Vibration"**, which is seen through **Emotional** changes within the **Human Body**. Also note that **"Emotions"** are nothing but **Energy** in constant motion or **Energy in Motion** (E= Energy then Motion).

Question: What is Vibration?

Answer: According to the Merriam Webster's Colligate Dictionary **Vibration** means: The Action of **Vibrating**; the State of being **Vibrated** or in **Vibratory Motion**: as (1): **OSCILLATION**. A characteristic **Emanation**, **Aura**, or **Spirit** that infuses or vitalizes someone or something and that can be instinctively sensed or experienced. A distinctive **Emotional Atmosphere** capable of being sensed.

All things are created in **Tones of Octaves of Energy**, and **Vibration**, just as there are Octaves of Musical notes; there are also **Octaves of Tones**. Between the 4th and The 15th Octaves, the normal Ear can hear **Sound**. The Human ear is the most acoustic receiver in Humans environment, the average ear can detect sounds intensities as low as 10-16 watt per centimeter and can stand intensities up to about 10-4 watt per-centimeter before pain ensures this our **Ear Drums** are called the **Tympanic Membranes**.

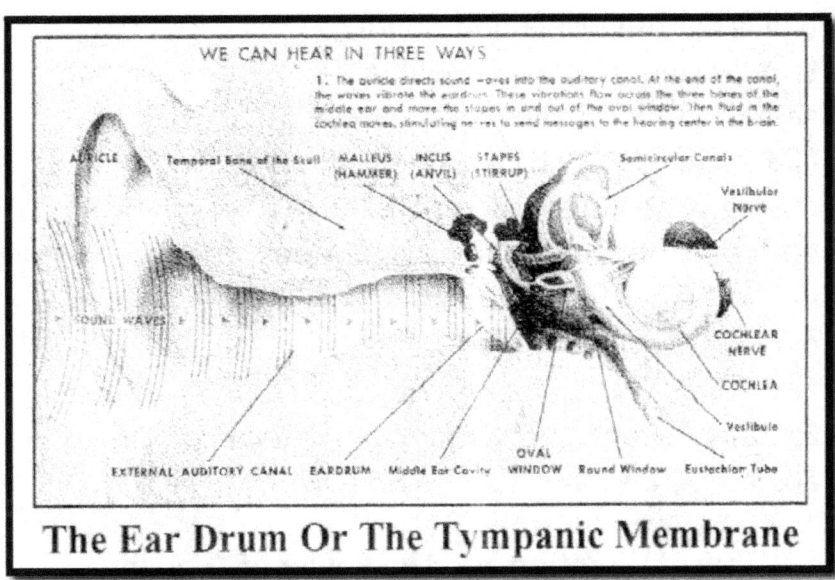

The Ear Drum Or The Tympanic Membrane

Now the 49th octave produces **Light, Cosmic Rays** manifest when energy increases between 72nd and 84th octaves levels. After cosmic ray octaves levels produce the **"Emotional Mental Wave"** to the sounds of **"HU"** which is **"Pure Green Light Energy"** (**Healing Energy**) (*refer to HU Are We? Master Key Series Vol. 11*) this is why when you **chant** this word **"HU"** with the right amount of input like **"1000"** times daily it will open up **Centers in the Brain** which causes all your Glands within your **Endocrine System** (**Chakras, Arushaat – Seats of Energy**) and bio organs to vibrate on the **Earth** (**PTAH-NUN**) frequency. The **HU** station to pick up the broadcasted **outformation** telling **Qatum** (**Melaninite**) **Beings** where to go, what to do, when the time is right, how to prepare for **the shift**. This will open your **Mind** (**KHU**), your **Heart** (**AB**), and give you a way to **Paradise**, which only means your **Etheric Body** (**AKH**) will be **"vibrating"** on the **cosmic tones** which raise your **vibration**. So it is the **Tongue** (T) those tones **Vibrating** which will open the way to **Paradise**, causing your **Heart** (H) and **Mind** (M) to open. Now you see the **Sacred Esoteric** meaning of **H.T.M** or **Heliopolis, Thebes,** and **Memphis** (*The HEART-The TONGUE- The MIND*).

Law of Vibration and The Power of Thought

"Nothing rests; everything moves; everything Vibrates in The ALL (PAA PAUT), yet movement is not ALL."

The great Third Egyptian Principle–**the Principle of Vibration**–embodies the truth that **Motion** is manifest in everything in the Universe–that nothing is at rest–that everything **moves**, **vibrates**, and **circles**. This Egyptian Principle was recognized by some of the early Greek philosophers who embodied it in their systems and gave birth to all of their thought of movement. But, then, for centuries it was lost sight of by the thinkers outside of the Egyptian ranks. But in this Century physical science re-discovered the truth and the 21st Century scientific discoveries have added additional proof of the correctness and truth of this centuries-old Egyptian doctrine.

The Egyptian Teachings are that not only is everything in constant **movement** and **vibration**, but that the "**differences**" between the **various manifestations** of the **universal power** are due entirely to the **varying rate** and **mode of vibrations**. Not only this, but that even THE ALL, in itself, manifests a **constant vibration** of such an **infinite degree of intensity** and rapid motion that it may be practically considered as at rest, the teachers directing the attention of the students to the fact that even on the physical plane a rapidly moving object (such as a revolving wheel) seems to be at rest. The Teachings are to the effect that **Spirit** is at one end of the **Pole of Vibration**, the other Pole being certain extremely **gross forms of Matter**. Between these two poles are millions upon millions of different **Rates** and **Modes of Vibration**.

Modern Science has proven that all that we call **Matter** and **Energy** are but "**Modes of Vibratory Motion**," and some of the more advanced scientists are rapidly moving toward the positions of the **Spiritualists** who hold that the phenomena of **Mind** are likewise **Modes of Vibration** or **Motion**. Let us see what **Science** has to say regarding the question of **Vibrations** in **Matter** and **Energy**.

In the first place, science teaches that all **Matter** manifests, in some degree, the **Vibrations** arising from temperature or heat. Be an object cold or hot—both being but degrees of the same things—it manifests certain heat **Vibrations**, and in that sense is in **Motion** and **Vibration**.

Then all particles of **Matter** are in circular movement (**Nutation** or **Nutational Rate**), from corpuscle to Suns. The planets revolve around suns, and many of them turn on their axis. The Suns move around greater central points (**vortexes**), and these move around still greater, and so on, as infinitum. The molecules of which the particular kinds of **Matter** are composed are in a state of constant **Vibration** and

Movement around each other and against each other. The molecules are composed of **Atoms**, which, likewise, are in a **State of Constant Movement** and **Vibration**. The **Atoms** are composed of Corpuscles, sometimes called "**Electrons**," "**ions**," on to **Quarks** and **Zeles**, which also are in a **State of Rapid Motion**, revolving around each other, and which manifest a very rapid state and **Mode of Vibration**. And, so we see that all forms of **Matter Manifest Vibration**, in accordance with the **Egyptian Principle of Vibration**. And so it is with the various forms of **Energy**. Science teaches that **Light**, **Heat**, **Magnetism** and **Electricity** are but forms of **Vibratory Motion** connected in some way with, and probably emanating from the **Ether**.

Science does now attempt to explain the nature of the phenomena known as **Cohesion**, which is the Principle of **Molecular Attraction** (**Law of Attraction**); nor Chemical Affinity, which is the **Principle of Atomic Attraction**; now they overstand Gravitation which was the greatest mystery of the three, which is the **Principle of Attraction** by which every particle or mass of **Matter** is bound to every other particle or mass. They found that the lightest Atom Hydrogen was itself not the smallest thing in existence but it was composed of even smaller things called **Quarks**, and even smaller **Zeles**, or **Zedes**.

These three forms of Energy now overstood by science, the writers incline to the opinion that these too are manifestations of some form of **Vibratory Energy**, a fact which the Egyptians have held and taught for ages past. The **Universal Ether**, which is postulated by science without its nature being overstood clearly, is held by the Egyptians to be but a higher manifestation of that which is erroneously called **Matter**–that is to say, **Matter** at a higher degree of **Vibration**–and is called by them "**The Ethereal Substance**."

The Egyptians teach that this **Ethereal Substance** is of extreme tenuity and elasticity, and pervades **Universal Space**, serving as a medium of **transmission of waves of vibratory energy**, such as **Heat**, **Light**, **Electricity**, **Magnetism**, etc. The Teachings are that The Ethereal Substance is a connecting link between the forms of **Vibratory Energy** known as "**Matter**" on the one hand, and "**Energy or Force**" on the other; and also that it manifests a **Degree of Vibration**, in rate and mode, entirely its own. Scientists have offered the illustration of a rapidly moving wheel, top, or cylinder, to show the effects of increasing **Rates of Vibration**. The illustration supposes a wheel, top, or revolving cylinder, running at a low rate of speed–we will call this revolving thing "the object" in following out the illustration. Let us suppose the object moving slowly. It may be seen readily, but no sound of its movement reaches the ear. The speed is gradually increased. In a few moments its movement becomes so rapid that a deep growl or low note may be heard. Then as the rate is increased the note rises one in the musical scale.

Then, the motion being still further increased; the next highest note is distinguished. Then, one after another, all the notes of the musical scale appear, rising higher and higher as the motion is increased. Finally when the motions have reached a certain rate the final note perceptible to human ears is reached and the shrill, piercing shriek dies away, and silence follows. No sound is heard from the revolving object, the rate of motion being so high that the human ear cannot register the vibrations. Then comes the **perception** of rising degrees of Heat. Then after quite a time the Eye catches a glimpse of the object becoming a dull dark reddish color. As the rate increases, the red becomes brighter. Then as the speed is increased, the red melts into an orange. Then the orange melts into a yellow. Then follow, successively, the shades of green, blue, indigo, and finally violet, as the rate of speed increases. Then the violet shades away, and all color disappears, the human Eye not being able to register them. But there are invisible rays emanating from the revolving object, the rays that are used in photographing, and other subtle rays of light. Then begin to manifest the peculiar rays known as the "X Rays," etc., as the constitution of the object changes.

Electricity and **Magnetism** (**Electromagnetism**) are emitted when the appropriate **Rate of Vibration** is attained. When the object reaches a certain **Rate of Vibration** its **Molecules** disintegrate, and resolve themselves into the original **Elements** or **Atoms**. Then the **Atoms**, following the **Principle of Vibration**, are separated into the countless corpuscles of which they are composed. And finally, even the corpuscles disappear and the object may be said to be composed of The **Ethereal Substance**, **Quarks**, or **Zeles** (**Zedes**). Science does not dare to follow the illustration further, but the **Egyptians** teach that if the **Vibrations** be continually increased the object would mount up the successive **States of Manifestation** and would in turn manifest the various **Mental Stages**, and then on **Spiritward**, until it would finally re-enter **THE ALL** (**PAA PAUT**), which is **Absolute Spirit** with the power to create or grow **Matter** out of itself. The "object," however, would have ceased to be an "object" long before the **Stage** of **Ethereal Substance** was reached, but otherwise the illustration is correct inasmuch as it shows the effect of constantly increased **Rates** and **Modes of Vibration**. It must be remembered, in the above illustration, that at the stages at which the "object" throws off **Vibrations** of **Light, Heat**, etc., it is not actually "resolved" into those forms of energy (which are much higher in the scale), but simply that it reaches a **Degree of Vibration** in which those forms of **Energy are Liberated**, in a degree, from the confining influences of its **Molecules, Atoms** and **Corpuscles**, as the case may be. Just as **Heat** escapes the flames.

These forms of **Energy**, although much higher in the scale than matter, are imprisoned and confined in the material combinations, by reason of the energies manifesting through, and using material forms, but thus becoming entangled and confined in their creations of material forms, which, to an extent, is true of all creations. The creating force becoming involved in its creation, fire to Heat, Water to wet, and Ice to cold. But the Egyptian Teachings go much further than do those of modern science. They teach that all **Manifestation of Thought, Emotion, Reason, Will** or **Desire**, or any **Mental State** or **Condition**, are accompanied by **Vibrations**, a portion of which are thrown off and which tend to affect the **Minds** of other persons by "**induction**." This is the **Principle** which produces the phenomena of "**telepathy**";

Mental Influence and other forms of the action and **Power of Mind over Matter**, with which the general public is rapidly becoming acquainted, owing to the wide dissemination of Egyptian mystery knowledge by the various schools, thought along these lines at this time. <u>**This is the New Millennium Teaching.**</u>

Every **Thought**, **Emotion** or **Mental State** has its corresponding **Rate** and **Mode** of **Vibration**. And by an effort of the will of the person, or of other persons, these mental states may be reproduced, just as a musical tone may be reproduced by causing an instrument to **Vibrate** at a certain rate–just as color may be reproduced in the same way.

By knowledge of the **Principle of Vibration**, as applied to **Mental Phenomena**, one may **Polarize** his **Mind** at any degree he wishes, thus gaining a perfect control over his **Mental States**, **Moods**, etc. In the same way they may affect the Minds of others, producing the desired **Mental States** in them. In short, be may be able to produce on the **Mental Plane** that which science produces on the Physical Plane–namely, "**Vibrations at Will.**" This power of course may be acquired only by the proper instruction, exercises, practice, etc. the Science being that of **Mental Transmutation**, one of the branches of the Egyptian Art. A little reflection on what we have said will show the student that the **Principle of Vibration** underlies the wonderful phenomena of the **power** manifested by the **Masters** and **Adepts**, who are able to apparently set aside the **Laws of Nature**, but who, in reality, are simply using one law against another; one **Principle** against others; and who accomplish their results by changing the **Vibrations** of material objects, or forms of **Energy**, and thus perform what are commonly called "**Miracles.**" As one of the old **Egyptian** writers has truly said: "He who overstands the **Principle of Vibration**, has grasped the scepter of Power, **Sekhem.**"

"**The Fourth Dimension** called <u>**The Dimension of Vibration**</u>" = Living in Spirit and Truth (**Intuitive Knowledge or Knowing**) a **Realm** or **Dimension** where there will be no more **Secrets**, where People and things will be seen and exposed for truly who and what they are.

The Vibrational Power of Thought

We must come to realize with our true inner eyes (Spiritual – 3rd Eye) that thought is a **Force** a manifestation of **Cosmic Energy** having **Magnetic** like power of **Attraction**. When we think we send out vibrations of a fine ethereal substance, which are as real as the **Vibrations** manifesting **heat**, **light**, **electricity**, and **magnetism**. These **Mental Vibrations** are not evident to our five senses are no proof that they do not exist. Just look at it this way, a powerful magnet will send out vibrations and exert a force sufficient to attract to itself a piece of steel weighing a hundred pounds, but we can neither see, taste, smell, hear nor feel the mighty force.

These **Thought Vibrations**, likewise, cannot be seen, tasted, smelled, heard nor felt in ordinary way; although it is true there are recorded cases where People peculiarly sensitive to **Psychic** impressions who have received powerful **Thought Waves**, and many of us can testify that we have felt the **Thought Vibrations** of others, both while in the presence of certain people and at a distance. People are becoming more and more sensitive to **Thought Vibrations**. In this day and time our thoughts and visions must be crystal clear, **Positive** and aligned with the will of our **Natural plan** and **Purpose** according to **Nature**. We are now living in a great era of time called the *"Solar Cycle of RE"* (*refer to Solar Cycle of RE, Master Key vol.2*) where the **veil of Isis** (**ASET**) is being lifted off the world, the Cosmic Veil of the Illusionary world of substance and the real world of ethereal matter. We have passed through the age of **Physical**, to the age of **Air**, **9Ether**, **Positive Thoughts** etc...We will soon really learn the *Power of Thought Vibrations* and the *Power of the Law of*

Attraction as **Qatum** **(Melaninite)** Beings. **Thought** is the fastest moving force in the Universe, and **thought** waves travel faster than the speed of light.

Nine Ether – The Original Cosmic Creator

Question: What is Nine Ether?

Answer: **Nine Ether** is the <u>Original Creator of the Universe</u>. To overstand what we mean **Nine Ether** can be seen as **Conscious** and **Conscience** Gases. As **Qatum (Melaninite) Beings** you are the Physical form of **Nine Ether**.

According the Merriam Webster's Dictionary **Conscious** means: **Perceiving**, apprehending, or noticing with a degree of "**controlled thought**" or "**Observation**": AWAKE: marked by strong "**feelings**" or **notions. AWARE**

KEYWORDS: **Perceiving** (Perception): **Observation** (TO OBSERVE as the Observer): **Awake** (Fully Conscious): **Aware** (Knowing)

According to the Merriam Webster's Dictionary the word **Conscience** Means: The "**sense**" or "**consciousness**" of the moral goodness or blameworthiness of one's own conduct, intentions, or character together with a feeling of obligation to do right or be good.

Keywords: **SENSE (INNER KNOWING – INTUTION, INTUTIVNESS), CONSCIOUSNESS (AWARENESS)**

So we as a **Qatum (Melaninite) Beings (NINE ETHER)** once we are back in our right mind fully using our **MIND POWER** thus **GOD POWER**, our **"Perception"** Level is at **full maximum** and we observe everything thus becoming the **"EYES"** of the Universe (ALL) as a witness, we start to live off our **inner feelings "intuition"** thus becoming truly **"AWARE"** of a World that most do not even know exist. In becoming what you truly are you have now become a **9ine dimensional being** that knows truly **Conscience**, **"Right"** from **"Wrong"** or **Aught** and **Naught**, meaning that which I **"Aught"** to do according to Nature (Neteru) and that which I **"naught"** to do according to the **Living laws of Almighty Eternal Nature (Neteru)**.

Nine Ether your true **Ancestral Spiritual (Etheric) Forces** in their **Etheric** form grew the Universe, which was the **primary creation** before the **lightest Atom** known as **HYDROGEN (H1)**. We must being to realize that energies existed as a form of energy existing in the form of Gases. **Nine Levels** of them existed from

Quarks to **Bi-Aps to Zedes/Zeles**, today referred to as **"Sub-Atomic" Particles (Energy)**, before the weight or the Sum of any weight registering as **"Nothingness"** yet still existing. This means being **"Lighter"** then the 1st form of detectable existence.

Let's look at **Nothingness** or the **State of Nothingness** as this, modern day *"Quantum Physicists"* have termed the world that exist beyond the **Atom** which is the world of **Quarks, Bi-Aps**, and **Zedes/Zeles** as **Nothingness**, meaning **Hydrogen Atom** these small minute particles of energy exist but they cannot be compared to anything on this side of H1 or **Hydrogen**. So the best word that was given was *"Nothing"* (*No.Thing*), and when you truly examine the word *"Nothing"* you see No – Thing or No. which is the abbreviation for the word *"Number"* and then *"Thing"*, the same with the word something or Sum –Thing, the *"SUM"* of a Thing. In Mathematics we learn that **Sum** is the totality, so the world existing below **Atoms** does not **Sum** up to or **ADD** up to any *"THING"* in Physical existence or Reality, yet it still *"Exist."*

Nine Ether is the combination of all existing Gases anywhere in Nature. Nothing can be as powerful as all existing Gases. So Nine Ether our true Ancestral forces performed Primary (1st) Creation and again this simply means that as Nine Ether forces called Etherians (Etheric Forces) they Grew the Universe. We as **Qatum** (**Melaninite**) Beings are part of the original growth of not only Earth (PTAH-NUN) but the entire Universe. So Qatum (Melanin) grew with Original Creation, to come along with a things means to grow with or within it. Nine Ether is 9 points from Ether 1 into darkness called "**Beyond –Beyond**", that pure tranquil, blissful state, and we **Qatum** (**Melaninite**) Beings manifest from point 1 in Hydrogen on into 9 Elements, and please know that the 8th Element being **Oxygen** for life the "DOT" in ancient TaMa-Rean language called "**Nagut**" Esoterically meaning "You stand in the Center of All Things". The Point of Origin of Things, first "**SUM**".

We as **Nine Ether** Beings (**PAA QATUM – The Melaninites**) utilize the forces that yield Energy versing Energy into one form called the Universe. In other words we always utilize the forces of the Universe to do our bidding. We always relied on the forces of Nature (Neteru) on and off the Planet thus "THE ALL in ALL", THINK...

Nine Ether is "**Positive Energy**" representing **Healing, Health**, and **the power to generate Life from within oneself** as a **Nine Etheric Being** (**Qatum – Melaninite**). **Nine (9) Ether** is the **Whole Light**

Spectrum of Electromagnetism, but the highest point of **Nine Ether** is **Tranquility, Peace**, and **Supreme Balancement** not Chaos. We as **Qatum (Melaninite Beings)** came down from the **Spirit Realm of Peace** and **Tranquility** and the farther we should drop down in this day and time emotionally is **"GREEN"** which represents **Healing, Health** and **Vitality**.

In Mecca today you have what is termed **Ka'aba** or **KAA –BAA** which is the Ancient **TaMa-Rean (Egyptian)** word for Spirit **(KAA)** and Soul **(BAA)**. At this **Ka'aba** you have what they call the **"Black Stone"** which is said to be the symbol of the Navel of the Planet Earth and the Flesh of **Adam** or in Arabic **"Adamah"** meaning a being from the Earth or **"Earthling"**.

In the **Esoteric Schools** of thought the symbol of the **"Black Stone"** has always been a symbol of focal point. In Ancient **TaMa-Re (Egypt)** the **Black Stone** was a symbol of the **Black Dot** the essence of **Human Beings** the **"Real True Self"** or **"Inner Being"** (The **BAA –SOUL**). In **Alchemy** which means the **"Black Sciences"**, **Black** meaning **Ether** or that **Peaceful State, Hidden forces of Nature**. This **Black Stone** was known as the **"Philosopher Stone"**, which when all base metals where transformed and transmuted you would reach a state of perfection, in your **Alchemical work.**

Even in this day and time Fraternal Societies like the Arabic Order of the Nobles of the Mystic Shrine would carry around a **Black Onyx Stone** as a symbol of those Protectors of the Sacred Secret, the Protectors of the Pyramids and other Sacred sites and Temples of the World.

This **Black Stone** really is the **Black Light Energy (UV –Ultra Violet Light)** of the full light Spectrum that our Minds are able to tap into as **Qatum (Melaninite) Beings** in this day and time. The **Black Stone** or **Black Dot** is also a symbol of **Nine Ether**. So the **Esoteric** or **Hidden Occult** meaning Islam YOU "ALLAH" was supposed to be taught through Mystical orders like the Sufi to make that **inner journey** pass the **Hydrogen** (H1) Atom (**ADAM** or **ADAMAH**), **pass the Quark** or the **4 Planes of Existence** which are **Solid – Plane of the Physical, Liquid – Plane of Forces, Gas – The Plane of Spirit** onto the **Fourth (4th) The Mental Plane –where your thoughts are formed** and find their origin.

In Islam you will find the **8 Pointed Star Symbol** of **The Mental Plane** meaning **"As Above so Below"**. So all this leads back to **Nine Ether**, the more you bring **Nine Ether Particles** into your Life, through **Positive**

Attitude, Positive Visualizations, Meditation, Prayer
and **Mental Focus** on a **"Green" Light Healing Force**
little seeming Miracles will begin to happen in your
life and you will start having a full effect on the
Quantum Fields of Reality, thus you being the
shaper and molder of situations in your life and not
controlled by them. This is important because the
more you begin to resonate on **Nine Ether Healing
Life Energy,** then you move to a **Higher Vibration,** a
higher orbital Spin within your **Atoms** of your Body,
thus having an effect on your over all **Consciousness.**

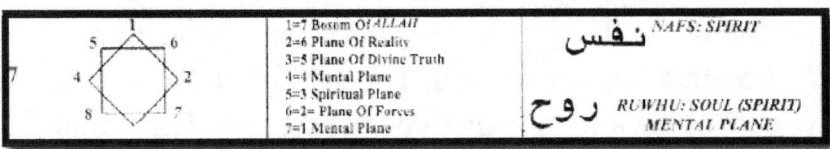

The Mental Plane - Doorway to ALL

We are made up of **Atoms** thus we were called in the
School of Religion **"Adam"** in Ancient **TaMa-Re
(Egypt) "Atum"** or **"Atom"** the 1st point of Light. We
as an **"ATOM"** are nothing but **Emotional Energy** or
Energy in Motion, and we are directed by the Mind
our **"Thought"** Power. The **Thoughts** we keep is what
we **attract** because **Thought Energy** is **Electric** and
Magnetic in Nature, so there are **Etheric Particles**
that **Attract** to us when we are thinking Negative,
depressive thoughts, which are Anti-Life and are the

Death Forces in Nature. The more you think negative the more you "**Attract**" negativity in your life.

Now get this! **Nine Ether** deals with **Health, Healing**, and **Positive** in Nature, so when you begin to think "**Thoughts**" of this Nature you "**Attract**" to your Life giving **Electro-Magnetic Particles** that correspond to these Healing Life giving thoughts. You seek **Peace** and **Tranquility** in your Life, you seek "**Positivity**" in your Life, then get on the **Nine Ether wave length** by first simply starting to think more better quality thoughts and seeing how you as a **Qatum** (**Melaninite**) Being is linked to ALL and you can have a "**Positive**" effect on your Life and all those around you. You want a **Positive Successful Life**, think **Positive** and **Success**. Nine (9) Ether is also **Effective** (Action) **Fluid** (To Flow, Movement) **Fire** (Burning Passion, Liveliness of Imagination: "**Inspiration**"). Take action and move by inspiration and revelation. Expanded Consciousness and Awareness, only Conscious perception alters reality as we know it.

The Number Nine (9)

The Number **Nine (9)** is the symbol of the **Circle all encompassing, Mind of the Universal ALL**, thus in Qatum **(Melaninite) Beings** it is our **MIND POWER! Nine Ether** is Positive **"LIFE"** giving forces in Nature the Original Creator, **PURE LIFE! Nine Ether** also represents **"beyond Infinity"** Nine to the Ninth power of Nine (999), **9 = Black Skin, 9 = Black Mind, 9 = Black Soul. 9 = the Perfect Cycle where all "Realities"** meet.

The Number 9 is a powerful and symbolic number. The number Nine is symbolic of **Heaven, Hell**, and **Creation**. Nine is a number whose square root is three(3), and Nine is a number when multiplied reproduces the same figures from up to down and from down to up, and it equals itself.

(9) Nine the Supreme Number

- 0+9=9
- 1+8=9
- 2+7=9
- 3+6=9
- 4+5=9
- 5+4=9
- 6+3=9
- 7+2=9
- 8+1=9
- 9+0=9

3x3 = 9 – 9 Points of Self "The Black Dot"

3x3=9 Triple Darkness The "Root" of ALL Thinking

1. The State of Quarks, the First Degree of Darkness, First Degree of No.Thingness
2. The State of Bi-Aps, The Second Degree of Darkness
3. The State of Zeles/Zedes, the Third Degree of Darkness

As Qatum (Melaninite) Beings we are comprised of Nine Principles and parts of our Being that make us who we are. The 9 Principles that make up YOU:

1. **KAA** – The Spiritual You
2. **KHU** – The Mental
3. **KHAT** – The Body
4. **BAA** – Your Soul
5. **KHAYBET** – The Plasmatic (Blood)
6. **AKH** – The Etheric You
7. **HATI** – Your Physical Heart
8. **AB** – Your Spiritual Heart (Inner Heart)
9. **SEKHEM** – The Spark of Life

Nine (9) Ether, 9 points from Ether 1 into Darkness (**Spiritual Realm – Pure Ether**), that **peaceful, tranquil, supreme state of Mind (Supreme Consciousness – Supreme Balancement**), we **Qatum (Melaninite)** Beings **manifest** (Materialize) from point 1 IN Hydrogen on into 9 Elements, the 8th being oxygen for Life, the dot "Nagut" is the point of origin (**NUN, BLACK LIGHT**) of things, the first Sum.

Our **Nine Etheric Ancestral Forces** that still reside within us, for we are them and they are us, known in Ancient TaMa-Re as the **Sapzdetu (*Sedjet, Enneads*)** The Head of **PAA Neteru Shil Neteru (The Nature of Nature)**.

The Nine Sapzdetu are:

1. **Atum-Re**
2. **Shu**
3. **Tefnut**
4. **Geb**
5. **Nut**
6. **Asaru**
7. **Aset**
8. **Sutukh**
9. **Nebthet**

The **Nine Sedjet** symbolizing this side of **Hydrogen (H1)**, this side of visible things, the world of **Light**, **Motion**, **Vibration**, **Friction** and **Chaos**, Energy in Constant Motion or Emotion.

Nine Reason – The Nine Mind

Reason is the mental process of figuring out something. Reasoning is the mental process of finding the answer to a question or the solution to a problem. Coming to a conclusion based on "**Facts**" and "**Confirmations**". **Nine Reason is Mind Power, which is God Power**. The ability to **Mentally** reach back in the **Black Ethers** (**NUN**) of your **Mind** to re-connect and pull out greatness!

Question: Who is NUN?

Answer: In Ancient TaMa-Re **NUN** was one of the 8 Ancestral Forces of Nature called **Khemenu** (**Ogdoads**) coming out of the Dark Watery Abyss symbolic of the world of Ethers or Etheric Realm also called the Underworld or Duat.

8 Khemenu Represent Energies in Almighty Eternal Nature.

Masculine Principles

1. **Nun** = Primordial Waters, Deep Abyss, Etheric Realm.
2. **Heh** = Boundless
3. **Kek** = Supreme Darkness
4. **Amun** = Hidden

Feminine Principles

1. **Nunet** = Heaven
2. **Hehet** = That which has boundaries
3. **Keket** = Light
4. **Amunet** = That which manifest

Nun = Primordial Waters (Abyss) Etheric Realm, Water Symbol of Spirit (Life).

In this day and time **NUN** is the Energy of Original Creation (**Positive Energy — Life**) which you as a **Qatum (Melaninite)** Being is linked to, that will take you into the Hereafter. The Hereafter which is *Mental Resurrection*, true Mind Power, **GOD POWER** and Ruler ship by Nature, where you are living from inside out meaning you are living from inner prompts, inner knowing as we call it "**Intuition**" or "**Intuitive knowledge**" which is the voice of The ALL in ALL, where we go from Conscious to Subconscious onto Super conscious.

Question: What is <u>Intuition</u> and how important is <u>intuition</u> in this day and time?

Answer: According to Merriam Webster's Dictionary **Intuition** means: Quick and ready insight. The **Power** or **Faculty** of attaining direct knowledge or cognition without interference of rational thought.

The word **Intuit** means: **to Know, Sense**.

Key Words: POWER: FACULTY: KNOW: SENSE

From the above definitions you can see that, taking the time to develop your **Intuitive** side of your **Nature** will give you the **Power** or **Faculty** to know and sense all things good or bad, agreeable or disagreeable. Ask yourself how does an animal like a Dog after falling far away from their home know how to travel great distances to return back home to their masters. The reason being Animals use a faculty that we long lost thus it is said Human Beings have "**FALLEN**" away from Nature and the Natural way of life.

To be **intune** with **Nature** is to live according to the **Laws of Nature** in **NATURES** Cycles and **due seasons**. To be **intune** with **Nature** is to go with the flow of

Nature, to be **intune**, in sync and in harmony with **Nature**. Your **Intuition** is the **Voice of the Subconscious** being, that deals with your **Emotions** and **Vibration** for as we now know, **Emotions** is nothing but "**Energy in Motion**". Energy in Motion causes ripples which produce "**Vibrations**". In this day and time Physicists are finding out what our Ancient TaMa-Reans (Egyptian) Ancestors knew all along that the **Subconscious Realm** occupies the **Fourth Dimension** which is the Dimension of "**Vibration**". We are linked to ALL, and this we must start to realize and embrace. **NUN** is the **Energy** or **MASTER KEY (☥) Ankh** that will open the door to the many Psychic Realms within our own being.

Our **Intuition** is very important in this day and time, your intuition which is the voice of your **Subconscious Mind** is linked to the **Plane of Force** or **Plane of Forces** which is the Realm the Forces of Nature operate. Planet Earth (**PTAH-NUN**) in this day and time is going through many changes what the Ancients call a "Shift", this "Shift" is a **shift in Dimensions** and **Polarity**, the shift that is NOW taking place will have a grand affect on the way all **Biological** life lives and is seen in the future. This means as Planet Earth (**PTAH-NUN**) is shifting so are we for we are part of this Planet we are the **Microcosm** of the **Macrocosm** (BIGGER PICTURE).

Planet Earth (**PTAH-NUN**) is now sending out **signals**, **thought waves**, and **emotional feelings**, for our Planet (**PTAH-NUN**) is alive; it breathes, thinks and feels. Planet Earth (**PTAH-NUN**) now currently is also going through birth pains, and a major transitions, we are in the middle of many Cycles. We are going through a *Polarity Shift* which last occurred 10,500 years ago recorded in ancient TaMa-Re and Sumerian culture as the "**DELUGE**", then you have the End of a 24,896 year Cycle which is an **Equinox**, then we are at a 50,000 year Cycle called a **Epoch**, so all this means the TIME is Truly NOW to tap into the Latent Powers and abilities of your Mind and realign your true self with Nature.

As **African People** worldwide and the **Children of the Ancient TaMa-Reans (Egyptian**s), the **Children of Nature**, the **Science of NUN** is our **true power** and **energy source known as NUPU (Sound Right Reason)**, the **DUAT** or **Etheric Realm** and your **Conscious Mind** is the **Key** to open the doors to your **Subconscious** which leads you into ALL. Your **Intuition** when aligned back with your true self, your inner being (**Soul**) which is the "**Emotional Energy**" linked to **Nature** will lead you in this day and time, you will begin to have an inner knowing, inner conviction as the **Children of Nature (Qatum Beings)**, you will know where to go and where to be to make this "**Shift**", this is truly a great day and time! (**Refer to PTAH-NUN 2012-2013 Book**)

Question: Who is the Ancient TaMa-Rean Deity (Neter) RE?

Answer: **RE to make/to dispose**, **RE** was a **manifestation** of the **Neter (Deity) Amun**. **AMUN-RE** or **AMON-RA**, who is also called **Amon-Re, Amun-Niu, Amen, Ameen**, and **Al Mu'min**, can be found in the New Testament in Revelation 3:14, and I quote:

*"And unto the Angel of the Church of the Laodiceans Write; these things saith **the Amen**, the faithful and true witness, the Beginning of the creation of God".*

This quote, as quoted from the **Christian Bible**, shows the definite article "**The**" before **Amen**. The Qur'an "Koran" also uses it as **Al Mu'min** (**Koran 23:1, 4:92**) in countless places. Booth of these __mythological religions__ are subtly giving praise to your **Ancestors**, TaMa-Rean Neter "**Egiptian Deity**" **AMUN, AMON**, who was the husband of **MUT**, ruler of the Heavenly skies.

His symbol is **PAA RE** "**THE SUN**", **the Fire of Nature** that causing skin cancer and eventual elimination of people that are **Melanin recessive**. As for **MUT**, the ruler of the heavenly skies and a reflection of **AMUN's light**; her symbol of course is **PAA AH "THE MOON"**. **AMUN** and **MUT** are symbolically **the Sun** and **the Moon**.

Amun is the origin and essence of all things and is the **inner most reality** of all Human Beings. **Amun** is the same **Life Force** (**9 Ether**) essence, which manifest through and from **THE SUN** (**PAA RE**).

RE is the **Divine, Universal Consciousness**, and the **Living Mind of Nature**. **RE = the Hidden Source of the Eternal, self originated, invisible, Eternal Being**, "**Father in Spirit**" (*Masculine Energy*). **RE= THE QUARK, the Father of Energy. Amun-RE** = the "**Hidden**" **Light**, within each person, **the inner you**, **the Soul**, Green Light onto Black Light energy, that is existing but most do not comprehend.

Now is your time as **Children of RE** or simply the **Children of the Black Light, Amun-RE** (**Hidden Light**), **RE also equals Ether. The Solar Cycle of Re** IS YOUR RE-CONNECTION with **TA-MU-NEFU-SET –HU.**(*Refer to the Solar Cycle of RE Master Key Vol.2*) That is **Ta** (Earth), **MU** (WATER), **NEFU** (AIR), and **SET** (SUN), all manifesting from **HU** (That creative force of Will).

Also the **hidden Esoteric** meaning of "**RE**" is the **Creative power, to make, Cosmic Consciousness,** and in Human Beings it is **Consciousness "the Aware Mind"**. Consciousness is the receiving capacity of the Soul (BAA) or individual intelligence. Another way of saying this is *"the ever burning light or first manifestation of the Universal Intelligence or Universal Soul."*

RA= the Visible Light (SUN)

Re= Energy Behind the Visible Light (SUN). Black Light (NUN or A-NUN-RE) "The Sun Behind the Sun". (*Refer to the Solar Cycle of RE, Master Key Vol.2*)

RE= Self begotten, Self Originated.

Align with the Black Dot within Self!

RE (Ether)

The Black Dot
Black Light
Green Dark
Soul Force

= 9 MIND
Eternal Mind

AMUN-RE
"The Hidden One"
The Hidden Light within you.

Time to Make the "Journey Within"

The TaMa-Rean Neter (Egyptian Deity) AMUN-RE (RA)

Question: What is Quantum Physics?

Answer: **Quantum Physics** is the <u>Physics of Infinite Possibilities</u>. **Quantum Physics** is the study of **Sub-Atomic energies** and **Particles** that are **"Sub"** meaning **"Below"** the **Atom** (**Hydrogen –H1**), the study of **"the All"** in things, thus **the ALL in ALL**. **Quantum Physics**, **Quantum Reality**, the level of full Conscious Awareness.

Question: Is there a Connection between Quantum Physics and Qatum (Melanin)?

Answer: Yes there is definitely a connection, through your **Melanin** which is a condensed form of **Light**, you as a **Qatum** (**Melaninite**) Being are connected to ALL, that which Scientist are calling **Quantum**. As **Qatum** (**Melaninite**) Beings we must learn the realities of Existence and reverse the process of "Self Deterioration", both Physically and Mentally, and with this confirmed fact become part of **ALL**, become part of the solution, "**the Cause**". We again must begin to realize that the Essence of our Being is that we exist and with that confirmation existence is as we are. Time is as we be! ALL is, ALL acts, ALL does, All things are a part of ALL on into Allness. The baby of Allness is found in the womb of **Quantum Physics**, Born from **Inside-Out** in Etheric Existence.

Question: Who are the Khemenu?

Answer: The **Khemenu** also known as the **Ogdoads** in **Ancient TaMa-Re (Egypt)** are the **8 Etheric Forces** from the dark watery abyss (**Etheric Realm**) leading to the 1st point of Light (**H1 – Hydrogen**). Before **Sound**, Before **Vibration**, Before **Color** and Before **Light, Motion** or **Emotion** (**Energy in Motion**). When these (**E-8**) **8 Etheric Forces** manifest on this side of **H1** (**Hydrogen**) they manifest as **sounds** as in the 1st **cosmic Sounds, Tones, Vibrations. Sounds** or **Tones** are grouped in sets of **Octaves** which is **8 complete Sounds, Vibrations** or **Tones**.

According to the Merriam Webster's Colligate Dictionary **Octave** means: (Middle Latin – *octava*, fr. L, FEM. Of *octavus* eighth, fr. *Octo* eight) the harmonic combination of two tones an octave apart. The interval between two frequencies (as in electromagnetic spectrum) having a ratio of 2 to 1.

An octave of musical notes are **CDEFGABC**, or "*Do, Ray, Me, Fa, So, La, Ti, Do*". In the visual Light Spectrum you have "**Red, Orange, Yellow, Green, Blue, Indigo and Violet**" 7 major colors with three primary. So sound is light and Light is sound because Sound causes friction, Vibration, thus produces a burning, things begin to move , go into motion thus you have Emotion (Energy in Motion).

Also note: that our **Genes (DNA)** <u>vibrates with the 8 Khemenu</u> (8 original Etheric Forces) as **Qatum (Melaninite) Beings**, as you remember we grew with Original creation thus we are still linked and always will be linked to the Source of all Creation **the ALL in ALL**.

8 Khemenu Represent Energies in Almighty Eternal Nature.

Masculine Principles
5. **Nun** = Primordial Waters, Deep Abyss, Etheric Realm.
6. **Heh** = Boundless
7. **Kek** = Supreme Darkness
8. **Amun** = Hidden

Feminine Principles
5. **Nunet** = Heaven
6. **Hehet** = That which has boundaries
7. **Keket** = Light
8. **Amunet** = That which manifest

8 Khemenu (Ogdoads) = **8 points** or **8 Ethers** once you step pass the **Hydrogen Atom** and begin to go back into the **Etheric World** the Realm of **Quarks, Biaps, Zedes/Zeles** etc..

As **Qatum (Melaninite) Beings** or **Children of the Sun (PAA RE)** each individual has their own **tone**, or

vibration that each person is **resonating** on. This is the note that brings you "**in tune**" with the **forces of Nature**. A good way of finding out your tone is by taking a collection of all the songs that make you feel exceptionally well, good and place you on a **high vibration** what most people term a "**Natural High**". You will begin to see that all of the songs you love have the same tone or the same major note which attracts you to this type of Music, thus it has your "note" in it. When listening to music that makes you feel good, happy and creates inspiration, you begin to start vibrating at a higher rate, and this causes your **Atoms** to move faster, which causes you to literally "**glow**" as **the Sun** (**PAA RE**). If you notice people with an abundance of **Melanin** (**African People –World Wide**) when we are exposed to **Sun Light**, because of our **Sun Heat Genes** spoken about earlier in this book, we begin to "**Glow**" to have this shine about our **Skin** "**TONE**." (Notice the word "**Tone**" as in **Sound, Octave, Vibration**)

Our **Skin** as **Qatum** (**Melaninite**) **Beings** has a "**tone**" which produces a "**Tone**", which is aligned with **Electromagnetic Frequencies**. Remember **Sound is Light**, and **Light is Sound**. If we were able to hook up a device and tap into the **Light** frequency of let's say **UV-Ray** (**Ultraviolet**) this Color and Frequency would produce a **Sound, Tone,** and **Vibration**. Now our

bodies again are **resonating** on our own **Tone**, **Sound**, and **Vibration**, **Frequency** that is linked to and in sync with the Universe (**the ALL in ALL**).

So once you as a **Qatum (Melaninite) Being** begin to live your true purpose and **"DESTINY"**, then you are back in sync with **Nature (Neter)** and the Universe (**The ALL in ALL**), you will see that your life seems to flow like "Harmonious Music"!

The Atom – A Tiny Universe

"The Atom is nothing more than a tiny Solar System charged with Electronic and Magnetic Power".

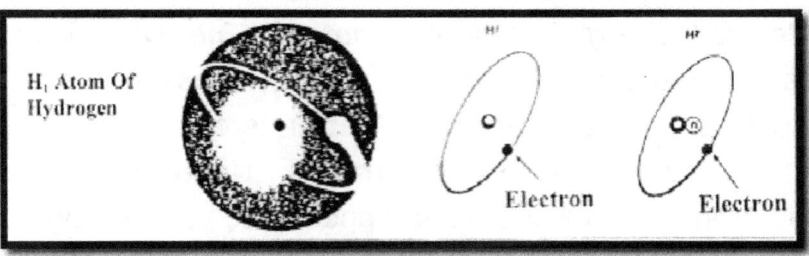

H₁ Atom Of Hydrogen — Electron — Electron

Question: What is an Atom?

Answer: Modern Physics are confirming what our **Ancient Egyptian Ancestors** knew all along. In the past 20 years, studies that have tried to find the smallest **Particle** or to explore the outer limits of space have come up with answers which support the

Ancient Egyptian teachings view of the cosmos and our relation with it as **Qatum** (**Melaninite**) Beings.

Science is discovering that the Universe is infinite in all directions, both at the Atomic (**Micro**) level and at the Planetary (**Macro**) level. Science is also finding that what they are calling "**Matter**", is not what it appears to be. In fact, studies that matter is 99.9% empty space surrounded by an idea, information, thought, and **consciousness**.

The "**Atom**" is said to be composed of a positively (+) charged "**Particle**" called a "**Proton**" and a **Particle** with no charge (N), called a "**Neutron**", in the center. These two particles are said to be surrounded by an **Electron** (*Energy in Motion, Emotional Energy*) which carry a **Negative** (-) charge and revolves around the **Nucleus** (Center, Axis). All Matter is found to be composed of the same **Protons**, **Neutrons** and **Electrons**. The difference in appearance comes from the different numbers of "**Particles**" in each "**ATOM**" (ATUM) and also from the combination of different atoms with varied combinations of the three particles, makes up all Persons, Places, and Things. So we must now learn to think past the **Atom**, meaning mentally go back past the **Atom**. Go past Person, Places, and Things, and think on a more

Quantum level, and learn to "**become the cause**" of the many effects of your life, in the **Positive**.

Further, it is known that **Electrons** have no weight and that there is a vast "empty space" between the **Protons** and the **Electrons** that circulate around them; also that there is "empty space" inside of the **Protons**, **Neutrons** and **Electrons**. Therefore, what we are seeing and touching by use of our senses is not all what it appears to be.

What we seems to perceive with our senses is in reality, only different aspects of the same substance. That is, when Energy "*vibrates*" at a *high speed* (*frequency*), it appears as a light (less dense, less weight) material such as gas or electricity. When it "*vibrates*" at a *lower speed*, it appears as a solid (dense material) object such as rocks or metal. The *Higher the Vibrations* are, the more subtle the "Material" will **appear** to be.

The *Slower the Vibrations* are the more solid, rigid and static it will **appear** to be. When matter vibrates at very high rates, it goes beyond the gaseous state; then matter appears as rays such as **Sun – rays** or **X-rays**. At Higher rates of Vibration, it would be so subtle that it could fit in between the "*empty spaces*" in the *slower vibrating matter*. It could pass through it or "*reside*" in it. This is the Subtle Realm of the

"**Spirit Body**" which inhabits the "**Physical Body**". The object of all spiritual movements is to "identify" one's consciousness, one's concept of who one is, with the "**subtlest reality**" rather than with the gross physical reality because the physical one is perishable and transient, whereas the subtlest one is transcendental and immortal. In fact, it is the "subtle" **Spirit** (**Ether**) from which "gross" Matter is created. For this reason, keeping a light and simple lifestyle which promotes **Higher Mental Vibrations**, a "light" diet and "light" thoughts are important in order to consciously experience more subtle vibrations of higher realities and to keep a harmonious flow in one's life.

The Physical world which appears to have defined boundaries is only an **illusion**. In reality, the world is one interrelated mass of **Atoms** and energy which is being "caused" to move and interact. The Ancient Mystical Philosophy of an all encompassing "force" that binds the Universe together (**Tachyon energy** – meaning to "**Tach**"- **On**) was espoused thousands of years ago in the **Egyptian Philosophy** of **Sekhemic Energy**, also called **Chi**, this force can be controlled through **Mental discipline**.

Modern science has now, based on scientific evidence, and have claim true the existence of a substance called "**Dark Matter**" which is described as an "unseen, unfelt substance that makes up to 99% of the Universe." This means that not only is the world one interrelated mass, but that it is a part of the greater mass called the "**Universe**". The Ancient philosophical idea is that the "Created" Universe really does not exist except as perceived through the "**Mind**" of the individual.

Ancient Mystical Philosophy states that the true essence of things can be seen by the liberated mind which sees what lies beyond the information given by the senses and that those whose minds are not liberated will experience the "physical" world as if it really "exists." For example: there is no blue sky. It only appears to be blue because of the limited human sense of vision.

Modern science now holds that nature and all phenomena occur because off an experimenters ability to conceptualize (**you reach into the spirit world and it reaches back**) the phenomena and to interpret it. Therefore, **the observer (*giving attention*) is part of the phenomena being observed**. Consequently, modern science now uses a new term for the experimenter. The new term is *Participant*. Thus, the experimenter is really a participant in the experiment because his or her consciousness conceives, determines, perceives, interprets and understands it.

No experiment or observed phenomena in nature can occur without someone to conceive that something is happening, determine that something is happening, perceive that something is happening, determine that something is happening, perceive that something is happening (through instruments or the senses), and finally to interpret what has happened and to overstand that interpretation. Therefore, the most recent theory in **modern Physics** is that **Matter**, that is to say **Creation**, is composed of not only **Energy** in varying degrees of density (**Vibration**), but that it is "**intelligent**", or it might be better overstood by saying that matter and energy are manifestations of Cosmic Intelligence (Consciousness).

What people take for granted is that they are solid beings when in actuality they are not! Humans are composed of **Molecules**, which are broken down to cells, which are further broken down into **Atoms**, which are broken down into **Elements** like **Hydrogen**. There is a formula a map to each Human Being. When you know the formula you can compose or De-compose at will, but you must have the map to bring any being from the other worlds here and when you get on the other side of **Hydrogen** you get into the **Ether Particles** or **Elements** like **Quarks, Biaps** and **Zeles**. It is certain forces within the Universe which hold these **Elements** together when you look at the word "together" you get "to" and "gather" which is referring to a mass or **Matter** that appears to be together but are not in actuality.

Tachyon Frequency is the Glue or Force in our Universe that hold **Atomic** and **Subatomic Energy**, together. This is the **force** which holds all **Matter** together. So what you appear to see as solid is not really solid but, are tiny particles being held together.

Therefore **Ether** is the **Element** of pure **Space** alone, and to move, you need **Soul (BAA)**, that is **Emotional Energy**, a reason to become.

Question: So when you start to Observer the Spirit (Etheric) world it looks back?

Answer: Yes this is true. The observer (*giving attention*) is part of the phenomena being observed. What you think you create, where your attention is there you are − and you are becoming that. What you meditate upon you become. **Attention** is power of energy (**e-motion**) in **action**. Life is a compelling force and your attention to a thing compels life to flow there. Where your attention is there you are −

what your attention is upon that you become. It is imperative to make effort control the attention. Through attention and governing of your feelings (**Emotion, Energy in motion**) you can become master of your world. The Power of Thought will become more and more important in these days and time. Now is the time to embrace a new type of Perception called "**Quantum Thinking**" (**Nine Reasoning**).

When you are dealing with **Quantum Thinking** you are dealing with **ALL** (**PAUT**), you are dealing with your connection with **The ALL** (**PAA PAUT**), "**MIND**" over "**MATTER**". The Nuwaupic word for **Melanin** is **Qatum**. The same as **Quantum** which is 4 dealing with our Ancient Principle of 1.**Atum-Re**, 2.**Atun-Re**, 3.**Amun- Re**, 4. **Anun-Re**, Being, the **Melaninites (Africans)** that have survived throughout all Cycles within the **Nature of Nature**, and we are the **Etheric** and **Physical Ascendants** of the **Descendants** or the beings that Decedent down here from **Triple Darkness** (**Ethereal Vibrations**) that have existed for 76 trillion years, in that Pure State of Blackness (**NUN**), Supreme Balancement.

The **Egiptians** were knows as **Alchemists**, that is **Chemists**, as you can see the names of the Deities and what is ascribed of the Deities and what is ascribed to them, they begin before the manifestation of **Nature** and **Matter** as you know it, which puts them into sub-atomic energies, as **Quarks (Spirit KAA), Bi-Aps (Soul BAA)** and **Zedes.**

Question: What is the Weight of a Zede?

Answer: On the <u>Elemental Chart</u> of **Matter** as you know it **"Zedes"** would have no sum, or weight, making them **nothingness** when weighed on the scale of sum, to something with weight by the **Law of Measurement.** Its center contains **Neutrons** and **Protons.** So, you can see where they get the concept of **Gods,** living in a **Spiritual Realm,** other than the **Physical realm** would be the **link** from the **Material Plane (Conscious Mind)** into the **Plane of Force (Subconscious Mind)** on to what becomes know as the **Spiritual plane (Emotions),** then **Mental (Super Conscious – Mental Reservoir, 4TH Dimension).** Notice, there are **Four Planes** and the word **Quarks,** or **Quantum,** is from the number **Four.** In Ancient **Tama-Re (Egypt)** you had the **4 Principles** or **Quadity** which is **Atum –Re, Atun-Re, Amun-Re** and **Anun-Re. The Mental** is the next level (**Orbital level**), Conscious level of Existence.

The School of **Mind** over **Matter,** where we are revolving to **Homo Spiritus, Mental Beings** were the 3rd **Dimension** interfaces with the 4th **Dimension.** We are now aligning our Minds with energies below the

Quark which is the **Father of energy (RE)**, to the **Biaps** which is the **Mother of Energy (Neith – Mitochondria DNA)**. It is time for the activation of **Supreme Melanin** which is **Neuromelanin**. We are going beyond the **Quark** now to **Biaps** also termed **Exotic Matter** or **Mesons**.

Question: What are Biaps or Exotic Mesons?

Answer: According to the Merriam Webster's Collegiate Dictionary **Mesons (Biaps)** are: any of a group of **fundamental particles** (as **Pion and Kaon**) made up of a **Quark** and an **Anti Quark** that are subject to the **Strong Force (Strong Nuclear Force)** and have **Zero (Zero Time Reference)** or an integer number of **Quantum** units of spin.

Mesons, the hidden meaning of the word **Masons** (**Ma Son of Re**), **Ma** = **Mother**, also **M=Mitochondria DNA** and **Other** you get the word "**Mother**", then you have **Son = Boy or Sun (Re)** – "**My Son of Re.**"

Ma is short for the **Sumerian Goddess Mami, Mammi, Mam –Mee or Mammitum**, The mother **Deity** or **Goddess** who procreated all **Mortals**. Taken from **Maat**, or **M'aat**, the Egyptian Mother "**Goddess of Justice**", the Mother of Justice and Order (**Order coming out of Chaos**). **Ma'at** is the **Netert, Guardian**

or **Supreme Being of Order**, again Order out of Chaos, so you have Order (**9Ether Life Cycle**) out of Chaos (**6ether Death Cycle**).

The *"Solar Cycle of Re"* (RA) (**The Cycle of the Quark – Quantum, Quantum Leap, and Mind Power**) comes back around (**Revolve, the Revolutionary Cycle**) with its **Zero Time Reference**, "**Now**" is the Time (**Mesons, Exotic Matter which is still Matter**).

Question: Is that where Religions get their Spiritual spook world from?

Answer: Yes, they refer to them as **plasma**, **plasmatic energy**, and **exmo-plasma**. The word **Plasma** is from the New Latin, from Late Latin, "**Image**", "**figure**", from Greek , from Plassein, "**To Mold**, " where **Energies** and **Gases** (**9 ether or 6 ether**) are Molded into images called **Etherians**.

We must begin to perceive (*Perception*) on a **Quantum** level that is "**Super Consciousness**", to be able to follow each cell in your body if you wanted to. These new generation of **Physicists** beginning with **Albert Einstein** have developed a "**New Physics.**" They now know that matter, that is, everything which can be perceived with our senses, including our bodies, is an "**ILLUSION**" meaning it

"**Alludes**" to something else of a higher finer quality within nature.

When look at **Matter** the way it truly is, we would see structures that appear as small **Planets** and **Moons** circling them at lighting speeds. **Matter** is made up **Atoms** and when viewing the structure of the **Atom** you will see that an **Atom** is not even solid but is made up of even smaller **energy particles**. The **Atom** is said to be composed of a positively (+) charged "**Particle**" called a "**Proton**" and a particle with no charge (N) a "**Neutron**," in the center. These two particles are said to be surrounded by an **Electron** which carries a negative (-) charge and revolves around the **Nucleus.**

So as you can see even the most solid looking structures are really moving; everything is in perpetual motion, nothing rests. Further, we would see that matter seems to come out of nowhere (**Now-Here**) and then goes back into "**Nowhere-ness.**" (**Now –Here**), because we are not going anywhere, we are not coming from, or going to, we are **NOW beings**, ALL in ALL. As all "**Matter**" is composed of the same "stuff," the different objects we see in the world are merely different combinations of the same material substance common to all things; this is what is meant by an illusion or **appearance of multiplicity** and **variety**. These "New Physics" say that "**Matter**" is nothing more than **Energy** (E-motion, Energy in Motion).

Particle accelerator experiments attempted to break down **Atoms** into smaller units by colliding them at great speeds. Scientists found that when a positively charged **Proton** (**Matter**) and a negatively charged **Proton** (**Anti-Matter**) are crashed together, particles turned into **Energy** (**Wave Patterns**) and then back to **Matter** again. **Energy** and **Matter** are therefore, interchangeable. This interchangeability of **Matter** and **Energy** is represented in the famous formula **E=mc2** by **Einstein** who initially developed this theory mentally (without experimentation). Therefore, even the most solid looking objects are in reality Energy in motion (**Emotion**) at different vibratory rates (**Vibrations**). **Atoms** (**Energy**) come together to create **Molecules**, and **Molecules** form objects. If **Matter** is in reality **Energy**, then what holds it together and causes it to appear as the varied "**Physical**" objects of the universe? **Matter /Energy** is held together by **Consciousness**. **Consciousness** is the underlying support of all things in the **Universe**. **Matter** cannot exist without your *"Consciousness Awareness"* to give it form and to be the **Perceiver** of its existence, because **Matter** is only an **illusion** projected by the **Conscious Perceiver** (**Awareness**) who uses **Sensory Organs (Sight, Hearing, Tasting etc...)** to perceive with, and a **Mind** to interpret that which is perceived. **Atoms** become **molecules** and **molecules** compose all **matter**.

Further, modern science has discovered that even objects of the world which appears to be separate, such as human beings, is in reality "*exchanging pieces of each other*" on a continuous basis. That is to say, every time we breathe out we are expelling **Atoms (ATUM)** and **Molecules** from our internal organs. Therefore, every time we breathe, we are sharing pieces of our bodies with other people and with the environment.

Another interesting thing to take note of is around every Human being there is what is called his or her "**Aura – An Electromagnetic Field**". Human beings are a Planet as well you are a living organism "**entity**" moving through out Space and in turn you as each Human being are a Universe of your own. You have millions up millions of micro organisms, living on and within you. Humans are the **Microcosm** to the **Macrocosm**. As above So Below, is a Universal **Sacred Egyptian Principle**.

Human's beings **Aura** is his or her **atmosphere** around there body (**Planet**) they exert influence or affects on their environment, some pleasant, some not so pleasant. Just like the Planet Earth (**PTAH-NUN**) has an atmosphere, so does the Human body called your **Aura**. Just like Planet Earth's (**PTAH-NUN**) atmosphere emits mood changes manifested by weather or in the form of weather, so does the human body. We call it **Emotions** (**Energy in Motion, Movement**) or **Electromagnetic Field**, which emit colors, vibrations, and tones, rhythms etc.

Your **Aura** is charged by your "**Mental Thought Waves**", the foods we eat affect our mental thinking capabilities. Human beings **Aura** are linked to the Planet Earth (**PTAH-NUN**) and everyone and everything around you. You are linked to the stars because you are linked to the Earth (**PTAH-NUN**), which is linked to a grander solar system "**the Milky Way**" (**where you will find the Central Sun called Central Vortex**) which is linked to other **Galaxies**.

Humans are linked to and are influenced by their environment; they are linked to the plants for plants have an **Aura**, Animals have an **Aura**, Trees, Bugs, Insects, Dirt, Water, All emit an **Aura**, a **Mood**, and a **Vibration**.

Now Scientist have learned that what they are calling **Matter** is not the final stage, and what they calling **Atomic Energy** is not the final stage. Once they found out that an **Atom**, a **Hydrogen Atom** which they have claimed is the lightest sum of energy for the last 100 years. Once Scientist found out that the **Atom** wasn't the lightest and that there where **Quarks**, and then after that **Zeles**. Always through school when they taught us about **cells**, they would always compare a cell to an **Atom** and an **Atom** to a cell. They would teach that an **Atom** has a **Nucleus**, and it has **Electrons** and **Protons**, moving around it, just like the **cells** in the <u>Human Body</u>.

So when Scientist became aware of the fact that **Hydrogen** which they thought was the lightest **Atom** is not that there is invisible **Matter** know as "**Dark Matter**", **Matter** that does not have a sum, by that I mean the word we use "**Something**" really comes out **S-U-M Thing**, the "**SUM**" of the "**Thing**", "**SUMTHING**" what it amounts to. So Scientist say **Hydrogen** is the lightest **Atom**, or the lightest thing in existence, once they found **Quarks** they have to alter, all the scales and all the weights be it metrics or whatever. Because there is no weight for **Quark**, there's no substance for the **Quark**, there's no density for the **Quark**, so it is actuality a "**Spiritual Thing**", they have confirmed exist when for years they couldn't prove that the spiritual world exist. The acceptance of an **Element** lighter than an **Atom** is admitting that there is some existence intelligent that is not perceivable by any equipment that they have made to date.

So with that being related to energy, now let's go to the Human cell so when they found that out then they must have also found out that beneath the lightest, and the smallest form of **Matter** , there must also be intelligent **Energy,** so they confirmed the **BAA (Soul)**, and the **KAA (Spirit)** and the **AKH (Ether)** of ancient Egypt, that there is a spiritual world, there is an **Etheric** world and intelligent abode that they are now tapping into with computers because they could not tap into it with the ordinary mathematicians mind because they use math to do this.

So what they call these new "**Spiritual Computers**", are **Computers** that are able to detect or sense intelligence not perceivable by the Human eye. They are looking into **exmoplasma** and the spiritual world. Now what the creators of these new types of computers don't know is the gates of the spiritual world, and when they open them, when they make this **magnetic link** the way we do with **electricity**. When you make a mistake and grab electricity, it has a thing, you grab it and it grabs you, and it decides to travel from its destination and include you in its path, and it zooms through your whole body!

When they bite into this spiritual world with these computers what Scientist don't want to admit is that when they grab it, it's gonna grab them, now do they know that happens? Yes, it's called the **Ouija Board**, and they have admitted over the years if you play with an **Ouija Board** you can open up a **Porthole (Vortex)** or a gate to another side where you will be plague with **Demons (Demonic – 6Ether Forces) for the rest of your life.**

No Jesus is not coming through the **Ouija Board** for you, however there will be people on the other side **disembodied souls (Six Etheric Gas Forces in Nature, Death Forces)** that will speak to you and pretend they're Jesus, and then tell you "see you have to go out and save the world," "you got to go out and kill all those people," because these **Spirits (6ether, 3ether Gas Forces)** take pride in having an effect on the spiritual world. These are people who died prematurely, died in hate crimes, died by brutal murders, and they are now **ethericaly** trapped in this realm with you, they are called **disembodied souls**. Oh you feel them, you feel them on

the back of your neck sometime, you feel them breathing, you feel sometime when you are walking, they step periodically into this realm by accident but you know what enhances their presents? **The adrenaline in your body**, the **FEAR!**

See a dog also picks it up, that's why they say a dog can tell you are afraid and will go after you. Also a Dog can hear sounds you don't hear, and see things on a level, they see **Spirits** you don't see so can **Cats** and that's why they use **Dogs** and **Cats** in **witchcraft**, because they have a link to both worlds. Now they have come to the reality that these things are no longer the twilight zone, it's real! And now they are able to weigh the **Soul**, they are able to detect through **Kirlian photography** the reality that if they clip off your finger, that something is still there, and they can pick it up with a camera, they can actually film it. Did you know that, it's called **Kirlian Photography**, look it up! They can remove your arm and put it under a light, and they will see your fingers and everything come right back even though there's no arm there. This is a fact today!

So they have to accept that there was something about that **Bible**, Something about that **Koran**, and something about that **Torah** which all came from the **Egyptian** writings you overstand, something about that religion that they got from those **Egyptians** your **Celestial** and **Physical Ancestors** and there connection with the stars and there talk about the Soul (**BAA**) and the Spirit (**KAA**) and the Etheric (**AKH**) you, the **BAA**, the **KAA**, and the **AKH**, or the **Rooakh**, or the **Ruwh**, in the Torah, or the Holy Ghost.

Question: So at what Point in Existence do you meet God or Allah or Re (Ra)?

Answer: You meet this **Energy** in the **Gas**! Because as you scale up the planes from the **Material Plane**, (**Solid**) to the **Plane of Force (Water, Energy, Water creates Current**) and after the **Solid** which is the Mask, **the Body**. The **Liquid** which is the **Water**, the

Blood then you have the **Gas**. So you have **The Physical Plane, The Plane of Force**, and **The Spiritual Plane**.

Now when you take those 3 Planes and overlay them in Degrees of Existence, which are known as The Plane of Mortals and Men (**Physical Plane**), the Plane of Force (**Angelic Forces, Beings**), The Spiritual Plane (**The Plane of Ethers**), Now you can say you where "There" which is "Here" when the word said exist! You where in an **Etheric State** which became this **Physical State**, you were there which is "Here" when the word was spoken "Exist", or "Be". You exist through that whole state, either going this way towards **Hydrogen** (H1) or that way stepping back toward the **Etheric (E1) World**. The **Physical Elements** or the **Etheric Elements** are all within **The ALL**.

The **State of Nothingness** is when a form of **Energy** is in "**transit**" or "**transition**" to another. Meaning when you on the Earth Plane, go back to the Lightest **Element** before **the Quark** was identified, you got back to **Hydrogen** (H1) you know that because they called it one, you know that because everything on the other side of one(1) is zero(0), non-existent, **Naught** (not here). **Quark** was existing in a **State of Nothingness (No.Thingness)**.

We had to become **"Conscious"** that we needed **Spiritual Guidance**. We needed to become aware that we need a **Spiritual Guide** to raise us up mentally from Person (1), Places (2), and Things (3) back to all or **Quantum thinking** (**Nine Reasoning**).

So we Humans started asking about Beings that vibrate on different degrees of **Ether**. When you say from (**H1**) meaning **Hydrogen** into **Quark**, **Quark** is going to extend, until they overstand **Biaps**, which is a **Bi-Aperture**, the **Mother of Energy**, in **DNA** its **Mitochondria DNA**. **Being** really means **density level**, density level is speaking about a process. To **"Densify"** is to go towards hardening. When they speak about the vortex, and the stream of vortex's that is above us, and they speak about it in the **density levels**, you would only identify with **Density**, from where you are, and how dense you are now.

Question: You Mentioned Mitochondria DNA, How Important is Mitochondria DNA in this day and time?

Answer: Oh it is very important in this day time. Being we are Qatum (Melaninite) Beings we are linked to ALL, meaning All Space, ALL Matter and ALL time on into infinite and beyond. Well as you have been reading you now should realize through your **Mind Power** you are able to literally think past the point of when you where conceived in the Womb of your Mother and tap into Higher forms of your DNA. Just by what we call **"Quantum Thinking"** or **"Nine Reasoning"** you can scale the infinite and boundless field of your own Personal DNA which is linked to ALL through your Hadus Qatum (Divine Melanin) and make a mental link with all your Great Ancestors, and receive inspirations, revelations and Supreme Guidance, for your life and all others around you. Remember **"YOU"** are the Master Key!

Now when dealing with **Mitochondria DNA** which on a **Molecular** level is now known as **Biaps** or **Mesons** which is the **"MOTHER"** of all energy, puts your whole being (**MIND, BODY, SPIRIT,** AND **SOUL**) on the channel of receiving from the very essence of **LIFE Forces (Nine Ether)** throughout the Boundless Universe which we call **"Infinite and Beyond"**.

Question: So what is Mitochondria DNA?

Answer: Well according to the Science World they have discovered that **Mitochondria DNA** is the oldest DNA found on the Planet Earth (PTAH-NUN) passed down from mother to child, and in reality from Mother to Daughter. This DNA comes from outside the Nucleus in a compartment of the cell called **"Mitochondrion"** which produces nearly all the Energy to keep the cell healthy and alive. This DNA does not get mixed with the father's DNA; instead, it is passed on unaltered from mother to daughter to granddaughter, and so on through the generations. Thus it is perfect to trace ancestral relations.

This discovery, by **Douglas Wallace** of **Emory University** in 1980s, led him to compare this **Mitochondria** of about 800 women. The surprising conclusion, which he announced at a scientific conference in July 1986 A.D., was that the **Mitochondria** in all of these Women must have all descended from a *"single female ancestor"*. The research was picked up by **Wesley Brown** of the **University of Michigan**, who suggested that determining the rate of natural mutation of **Mitochondria**, the length of time that had passed since this common ancestor was alive could be calculated.

Comparing the **Mitochondria** of 21 women from diverse geographical and racial backgrounds, he came to the conclusion that they owed their origin to **"a single Mitochondria Eve"** who had lived in **Africa** between 300,000 and 180,000 years ago. This led to the research of Eve. **Rebecca Cann** of the **University of California at Berkeley,** had obtained the placentas of 147 women of different races and geographical backgrounds who gave birth at San Francisco hospitals, she extracted and compared their **Mitochondria**. The conclusion was that they all had a common female ancestor who had lived between 300,000 to 150,000 years. The upper limit of 300,000 years, **paleo-anthropologists** noted, coincided with the fossil evidence for the time **Homo Sapiens** made his appearance. Another Scientist made a discovery in prior to them back in 1974 A.D. named **Dr. Donald A. Johnson**. Her name was "**Lucy**" found in **Hadari, Ethiopia** bones dating back **3 ½ Million years**.

The **Mitochondria DNA** is only given from **Mother** to **Daughter**; the **Male** species does not have any **Mitochondria DNA**. This is just further proof that the first person to walk on the planet was a **female**, an **African Woman**!

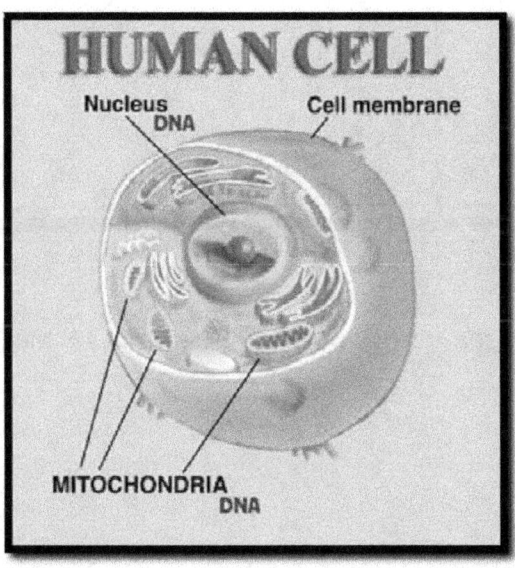

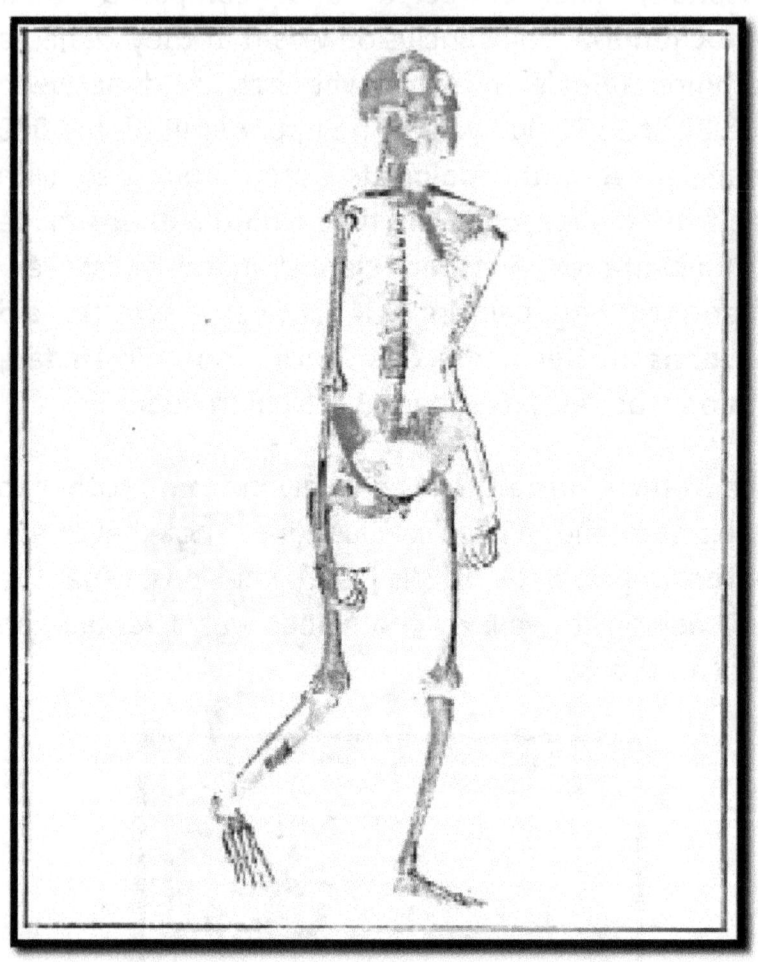

Lucy Oldest African Woman's Bones found. 3 ½ Million Years Old

Question: If Men do not have Mitochondria DNA, then how do Men tap into the Forces of Nature?

Answer: Men must realize that by nature you are "**Nature**" (**Neter**) as well, being **genetically** you do not inherit **Mitochondria DNA** does not mean you do not have direct access to **Nature** and the **Forces of Nature**. Being that all your Ancestors who are now on the other side of (**H1**) **Hydrogen** in the **Etheric Realms** of Existence residing within the **Plane of Forces** and **Higher Abodes** you have access to them via "**MIND POWER**".

You must remember us as **Qatum** (**Melaninite**) **Beings** men and women are all linked to **Nature** (**Neter**).Where a Man would have to implement certain Spiritual Daily Practices to stimulate and wake up dormant **Psychic Powers**, Women Naturally have the gift. Always during the **Solar Cycle of RE** (**The Sun Cycle**) Women always rise back to their position as **GODS** and "**Tribal Leaders**", known as a **Matriarchal Society**. It is during the **Moon Cycle** (**The Death Cycle of Nature**) That the **Feminine forces** in Nature in us and on Earth (**PTAH-NUN**) go to "Sleep" ONLY for the time where they will "**Spring**" or "**Revolve**" back around.

The **Cycle** we are in now The **Solar Cycle of RE** as **Qatum** (**Melaninite**) **Beings** we will all benefit and most importantly our **GODDESS**, the **Original Mothers** of the **Universe** will take her rightful position and the Men who have took the time to **transform** themselves back into the **GODS** they are and are open to the **Feminine forces** of Nature will ultimately benefit as well. Remember in the **Etheric** (Spirit) World there is no Gender just **ALL**.

Now in this day and time **Feminine Energy** is "**MOST**" needed in order to heal the Planet Earth (**PTAH-NUN**). It is the Feminine forces of Nature that we are dealing with in this day and time. Feminine Forces of Nature transcend Gender; those who are open to receive will Nature work through. We Children of the Sun (**PAA RE**) hold up our hands in reverence as healing energies are being sent down to recharge us and revitalize our **Supreme Melanin**.

DNA – Deoxyribonucleic Acid

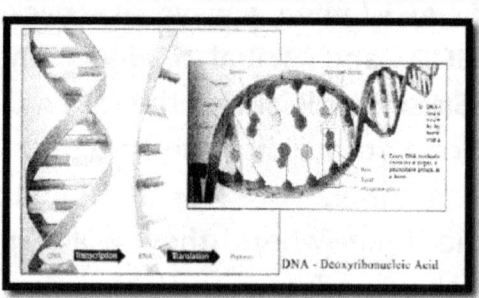

Question: What is DNA, and why is it important again?

Answer: DNA is so important because within your **DNA** through what is called "**Heredity**" there is a recorded past of everything that has ever happened to a person and there **family Lineage** (<u>bloodline</u>) throughout time, leading to the first groups of people on Earth. As we know, **all life started in Africa!**

With **DNA** testing many are able to find a direct **blood link to all Ancient** and **Great Rulers.** Many well trained **Master Minds** throughout history have been able to tap into this infinite resource dormant in their **DNA**, so "**NOW**" is your time to tap into your **Hidden Potential** and true **Powers of your Mind**, so that you can once again take your rightful place in the world. In order to know where you going you must know where you came from, **Sankofa!**

<u>**What are Molecules**</u>

Question: So what are Molecules?

Answer: A **Molecule** is the smallest **Particle** into which an element or a compound can be divided without changing its chemical and physical properties; a group of like or different atoms held together by chemical forces. The word **Molecule** comes from the French **Molecule**, from New Latin **molecula**, diminutive of Latin **moles**, meaning "mass". **Molecules** of different **Elements** may combine to form new compounds. All Substances are made up of very small **Particles** called **Molecules**. "**Molecules**" are always "**moving**". **Air Molecules** and **Water Molecules** are always moving, in constant motion. The speed at which **Molecules** move depends upon the **temperature** of the substance. In a warm substance, **Molecules** move faster than they would if the substance was cool. When the temperature of substance is raised, the **Molecules** move faster. The **Molecules** have more "**Kinetic**" **Energy** and can do more work.

Question: What is Kinetic Energy?

Answer: Great question, according to the Merriam Webster Collegiate Dictionary **Kinetic** is defined as: Pronunciation: \kə-ˈne-tik *also* kī-\Function: *adjective* **Etymology:** Greek *kinētikos,* from *kinētos,* from *kinein* Date: 1864
1 : of or relating to the motion of material bodies and the forces and energy associated therewith
2 a : active, lively **b** : dynamic, energizing
So as you can see, **Kinetic Energy** is Energy in Motion.

Question: Do 99 Elements relate to the 99 names of the Nature (Neteru)?

Answer: Yes the 99 names of **Nature (Neteru)** and the 99 **Elements** within **Nature (Neteru)** do relate. Each **Element** has a corresponding property, action or characteristic in the way they perform their job. The **99 Elements** also is the beginning of the Supernatural, and the Physical creation of the **Sun (PAA RE)**. Being that you **Qatum (Melaninite) beings** are all in ALL, Ultimately is most important to realize how they relate to you. WE are the most magnificent beings on the planet just by the dynamics of our Melanin alone. These are purely facts needed to be known in this day and time. With knowing these facts there comes a level of responsibility as the Daughters and Sons of Creation. As the daughters and Sons of Almighty and Infinite Nature, now is our time to shine!

What is Qatum (Melanin) Physics?

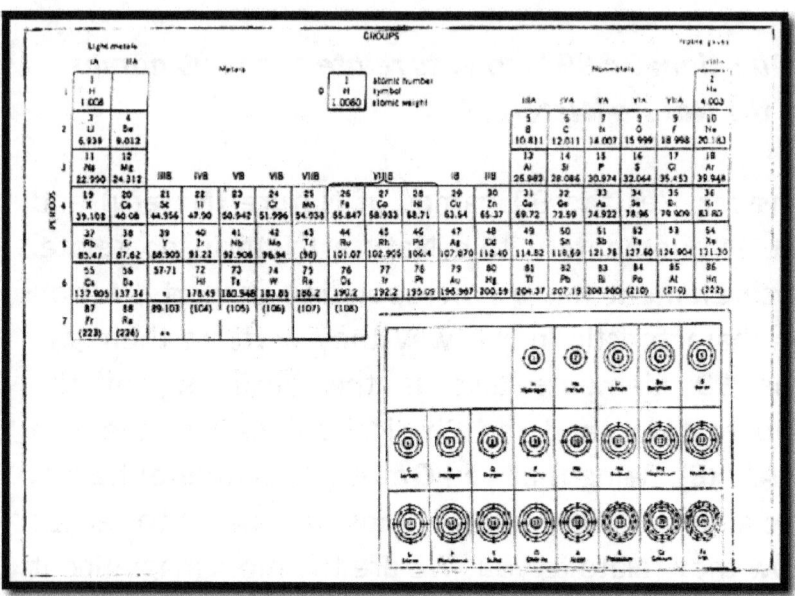

The 99 Elements within Nature

What is Qatum (Melanin) Physics?

1. **Atum-Re**	25. Mut	50. **Atun-Re**	75. Qebeh
2. Nun	26. Nefertum	51. Shai	76. Duamu-Tef
3. Nunet	27. Montu	52. Tehuti	77. Merit
4. Heh	28. Bebti	53. Seshat	78. Mafdet
5. Hehet	29. Hu	54. Tawaret	79. Mert Sekert
6. Kek	30. Hat-Har	55. Bast	80. Edju
7. Keket	31. Mehurt	56. Selket	81. Nekhebet
8. Amun	32. Khefri	57. Hah	82. Sekhat
9. Amunet	33. Tanen	58. Nehebka	83. Anukis
10. Hika	34. Raet	59. Sokar	84. Djet
11. Atum	35. Khentimentiu	60. Bait	85. Nebertcher
12. Shu	36. Heka	61. Aker	86. Ua
13. Tefnut	37. Sakhmet	62. Ini Herit	87. Uas
14. Geb	38. Anubu	63. I-M-Hotep	88. Anku
15. Nut	39. Khnum	64. Wapwawet	89. Afnuit
16. Aset	40. Khonsu	65. Sekhit Hetep	90. Satis
17. Asaru	41. Neith	66. Meresger	91. Sekhti
18. Nebthet	42. Bes	67. Sia	92. Mta-A'
19. Sutukh	43. Ptah	68. Gerhet	93. Anun-Re
20. Haru	44. Sia	69. Behtit	94. Imiut
21. Astennu	45. Heket	70. Kh-Nemtit	95. Imsety
22. Hapi	46. Amsu	71. Menqit	96. Qemamu
23. Anquet	47. Ma'at	72. Hem	97. Rehshef
24. Sobek	48. Serapis	73. Meskhenet	98. Mery
	49. Renentet	74. Meduty	99. **Amun-Re**

The 99 Names of Nature (Neteru)

Question: What is Matter?

Answer: Modern day **Physicists** have noted the fact that **Matter** is "**Thought Energy**", **vibrating** at a **Lower Frequency**. In Earlier times **Alchemists** performed their **transformations** of One **Mental** into another by using the same principles. In fact all "**Manifestations**" is merely the **Life Force** working at different rates and speeds of **Vibration** and the difference between one **Element** and another is merely its different **Frequency of Vibration**.

9Ether Beings (Qatum Beings - Melaninites) Africans, vibrating at a lower rate become decomposed into **Etheric beings**, but on the **Physical level** you **vibrate faster**. Even your **pigment** moves faster making you darker, and **Melanin** which also makes your skin darker, and the **organs** inside your very body are **vibrating** at different rates.

Question: What is Energy?

Answer: There is a great deal of **Energy** flowing freely throughout the **Universe**. This **Energy** is seen in the form of **Light**. In fact, light is the visible part of **Energy**, "**Electromagnetic Spectrum**." A **Spectrum** is a range of things related by certain characteristics. In this case, the **spectrum** is made up of **energy** that travels through space much like a wave that ripples in a pond. Scientifically the height of the wave is called the Amplitude or the height of its intensity. The distance between the peak of one wave and the peak of the next is called **wavelength** and the time per second that it takes a wave to pass through a specified point is the frequency.

That **Spectrum** is called "**Electro Magnetic**" because it has an **electric field** as well as a **magnetic field**. The **Electric Waves** of the **Electric Field** are identical in **amplitude**, **wavelength** and **frequency** as the **Magnetic Waves** of the **Magnetic Field**, hence, **Electromagnetic**. Thus we have the **Electromagnetic** nature of life.

When everyone was born, they received a "**Breath of Life**". That **Breath of Life** was in actuality a "**Breath of Light**". Not the light of a light bulb, but a light of life, giving, pouring forth forever emanating and forever penetrating. This light is what makes you a **living Soul (BAA)**, the pure green light essence.

This **light** existed **darkness** in a state of void before it became something or sum thing. **Life itself is a burning. Life** springs from an **acid base** and is perpetuated (continued) by that **acid base**. Of course, acid burn, so hence, life is a burning, like the acid in a car battery is the life of the car. The car is made active by gas burning and the person is made active by blood burning.

When a baby actively starts to breath the **Oxygen (air)** it breathes goes to the lungs, and **acid in the blood** in its **lungs**, and **acid in the blood** in its **lungs** burns the **Oxygen** the child breathes, and that burning is **natural electricity** called **Ether (Natural Ether)** which is **life**.

Electricity or **Electrical Energy** can be sent over long distances! **Thoughts** which are **Electrical** become things. They can be seen (**Light**) and Felt (**Heat, Cool,** etc...) **Thoughts** operate on the principle of **Electrical Energy**. **Electrical Energy** is of a special importance in energy changes. Most forms of **Energy** can be changed into **Electrical Energy**. **Electrical Energy**, in turn can be changed into almost any other form of **Energy**.

Magnetism:

A **Magnet** works off of the Principle of moving **Electric** charges. An **Electric** current produces a **Magnetic Field**. In **Atoms** there are **Electrons** moving around the **Nucleus** or Center of the **Atom**. In most **Atoms**, the **Electrons** spin in different directions and their **Magnetic fields** cancel each other. But in **Atoms** of a **Magnetic Element**, such as **Iron** and **Nickel**, the fields do not cancel each other. Once the fields get large enough so that they don't cancel each other, they are **dipole**, meaning pointing in different directions, then the **Magnet** becomes fully **magnetized**.

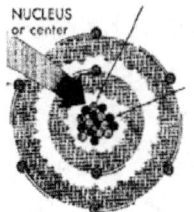
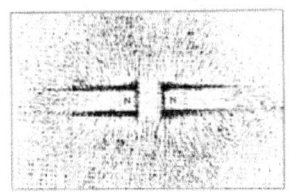

The Energy Fields Of A Magnet

Energy is Magnetic

One of the Laws of Energy is that Energy of a certain quality or **Vibration** tends to **Attract Energy** of the similar quality and **Vibration**. **Thoughts** and **feelings** have their own **Magnetic Energy** which **Attracts Energy** of a similar nature. We can see this principle at work, for instance, when we "accidentally run into someone we've just been thinking about, or the phone rings and the same just happened, the person we have been thinking about calls. Also we pick up a book which contains "**exactly**" the perfect information we need at the moment. **Energy** is **Vibrating** at different rates of speed and **Vibration** thus has different qualities, from finer and more refined to more dense. **Thought** is relatively, finer and of a **Higher Quality Vibration**, a light form of energy and therefore very quick and easy to change. **Matter** on this **Physical Plane of Existence** is relatively dense, compact energy, and therefore slower to move and change. Within matter there is a great variation as well. Living flesh is relatively fine, changes quickly and is easily affected by many things. A rock for example is vibrating at a slower rate thus appears to be a more denser form, slower to change and more difficult to affect. Yet, even a rock is eventually changed and affected by fine, **Light Energy of Water**, for example.

Emotional Energy

Energy is defined as the power by which anything acts effectively to move or change things or accomplishes any result, or "**power in active exercise**," or more simply, "**the capacity to perform work**." It divides energy into "**potential energy**" and "**kinetic energy**." Anything in motion has **kinetic energy**, a moving automobile, a falling rock, a breaking wave. **Potential energy** is sub-divided into "**available energy**" and "**diffuse energy**." To explain this we might say a spring that is used in a clock contains available energy, while a lump of charcoal contains diffuse energy, the difference being that the spring being used for the clock is ready for immediate work but the piece of charcoal must be burned before its energy is released in the form of heat.

All **life energy** is diffuse. It must be converted or transformed in some way in order to be put to work. Every **Human being** has an array of **Energy transformers** which enable him to draw upon the life energy which surrounds him and in which he lives, moves and has his being. Our **Ancient Egyptian Ancestors** termed these **transformers** as "**Arushaat**" (Arushaat) or **Energy Seats**. The **Ancient Hindu** or **Sanskrit** term for these **transformers** is **Chakras** or **Wheels**. This name comes from their appearance. To a person who's third eye is open and has inner vision these **Energy Seats** look like rapidly Spinning Wheels of different-colored **light**, coming from the **electromagnetic spectrum**.

The **raw life energy** is pure and without any distinguishing characteristics but when it is **transformed** into **Human Energy** by one of these **Energy Seats (Arushaat, Chakra)**, it takes on the quality of that particular **transformer** or **energy wheel**. The most fully opened and therefore most active **Energy Seat, Arusha** (Arushaat) or **Chakra** in average Humans is the **solar plexus (seat of the** subconscious mind), (*refer to The Solar Seat of RE Book*) whose chief product is **Emotional Energy**. In Consequence these People are **"Emotionally focused,"** which means that most of their actions are taken as the result of an **Emotional** impulse or urge. In order to open the **deeper gates** or **doorways of our Mind**, thus using our **full mental potentials** we must reestablish control over our **Emotional Energy** and how we use it.

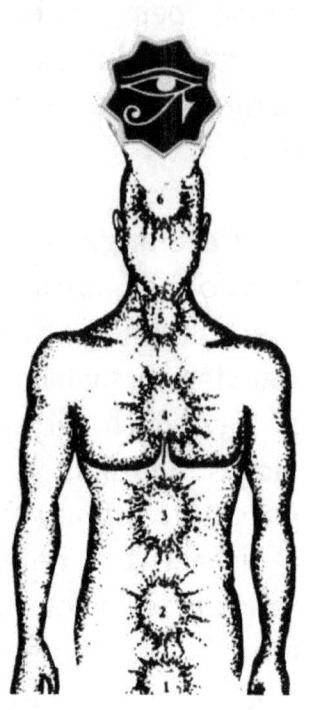

Question: What is Quantum Thinking (Nine Reasoning)?

Answer: Quantum Thinking is the ability to tap into your own great **Genetic Bloodline** and make contact with all your great **Ancestors** and let those great deities work with you and through you. Knowing that *infinite possibilities* exist in the Universe and being you are a **Qatum (Melanin) being** you are the linked to *infinite possibilities*. In this day and time *"Nothing"* *is impossible*. **Quantum Thinking** or **Nine Reasoning** is the process of learning to think past your **Conscious Awareness** to the more Subtle phases of your **Conscious Mind** known as the "**Subconscious**" and "**Super Conscious**" realms of thinking.

The symbol of the #9 (**Nine**) is the Circle which is a symbol for "**REASONING**", "**Completeness**" or "**Wholeness**". The process of **Reasoning** is figuring out a subject or a thing to its final conclusion, or to draw logical conclusions. **Nine Reasoning** is thinking and reasoning in the "**Positive**". Having the ability to **visualize** and sculpt your life for the "**Positive**". Learning and knowing you can have a **positive effect** on your environment. As a **Qatum (Melanin) being** you must "**know**" your great **potential**. Through your **Subconscious Minds** you all are linked to **ALL, and must learn to perceive (Perception) on a Quantum level that is "Super Consciousness".**

When a Person is **Positive,** he or she will indeed choose the life of life. When the person is **Positive,** he or she picks **Reason** who is ascending. **Nine Reasoning** is dedicated and obligated to finding the best answers to questions and also the best solutions to problems. **Nine Ether (9ether)** is a **Positive Force** within nature known in ancient times as "**NUPU**". **Nine Ether Forces** changes **Matter** for the better standards, values, and qualities in Nature. Our **Nine Ether Ancestral Forces** within Nature deal with **Truth, Justice, Equality** and **Freedom**, and these attributes are the basis of **Quantum Thinker (Reasonist)**, this is the types of thoughts we must formulate and keep on our mind in this day and time, for this is the **VIBRATION** the **Planet Earth (PTAH-NUN)** is vibrating on.

Nine Ether within **all Space, all Matter** and **all Time** is also called **"Positive Ether"**, and this **Positive Ether (the NINE)** controls and supports **the living** not **the dead** nor **mentally dead. Nine Ether** is the **"Positive" POWERS** in **Nature** and we are linked to these **Positive Powers** and **Forces in Nature** by Birth right through the very **Melanin** in your **Physical Make up!**

As a **Quantum Thinker (Reasonist)** and **Qatum (Melaninite)** Being you must align yourself with **Positive** and **Progressive** thinking at all times, for to do this is to be right in alignment with **Nature**, and **Natural Forces** in this time and season, which is **SOUND RIGHT REASON.**

You must begin to realize oh **Qatum (Melaninite) Children** that **SOUND RIGHT REASONING, NUPU, NINE REASON, ASCENDING REASON,** as well as the Best most practical **Mind**, uses all practical means possible including **investigation, examination,** parallels, **Geometric figures, evidence, examples, experience,** inspiration, **revelation,** and **experimentation** when possible, in order to reach binding conclusions.

SOUND RIGHT REASON is **NINE-ETHER (MIND POWER), concentrated (concentration) – profound-rational thoughts** and **irrefutable, conclusions** in the **Science of Natural Law** and **Order, Inspiration** and **Revelation, Knowledge, Overstanding,** and **Wisdom** called **Intelligence** and **Intellect. SOUND RIGHT REASON** is also known as **NUPU** and **NUPU** is plural in "**POWER**" and "**PURPOSE**".

In this day and time we must begin to realize as **Qatum (Melaninite) Beings** we have adopt and formulate a **Scientific Mind** and **Thought Process,** for **Science is Knowledge** and **Knowledge** is "**TO KNOW**". We must have the ability to distinguish from that which is Right to do for all and that which is wrong to do for all according to the **Laws of Nature.** Our Ancient Science which is **NUN** deals with **Reality, Reason,** and that which is **Practical** to do in this day and time. That which is right and correct to do according to the time and season we are in. Again **NUN** is our **Spiritual** and **Ethereal Science of NUPU** which are **Natures Forces** in the **Positive.** Now is the time as **Qatum Beings** to embrace a **Positive Mind** set in order to transform with the "**New (NU) Cycle**" changes affecting **PTAH-NUN**, the **Planet Earth**!

Question: What is The ALL verse ALL?

Answer: Well the best way to answer this question is by reviewing the great **Tehuti doctrine** that says:

"While All is in THE ALL, it is equally true that THE ALL is in All. To him who truly understands this truth hath come great knowledge."–The Wisdom of Tehuti

The ALL can be viewed as the **99 Elements** that make up all **Physical Matter** via All Person, Places, and Things on this side of **H1** (**Hydrogen**) and "**ALL**" comprises of all **Etheric Forces** that exist yet are hidden that H1 or Hydrogen and the 99 Elements manifest from, *"From the Unseen to the Seen!"*

Question: When you say ALL do you mean Completeness or Wholeness?

Answer: Yes, All Matter, All Space and all Time thus All In ALL. In Metaphysics "The All in ALL" is also viewed as "**Universal Consciousness.**"

Question: What is Universal Consciousness?

Answer: Universal Consciousness also termed **Universal Mind** or the **Mental Reservoir** is the home of **infinite and "ALL" possibilities**. It is through your **Subconscious Mind** that we are connected with **Universal Consciousness** and brought into relation with the infinite constructive, life giving forces of the Universe.

All **"Great Minds"** of the world past and present would agree that there is one **Principle** or **Consciousness** pervading the entire **Universe**, occupying all space, and being essentially the same in Kind at every point of its presence. It is all powerful, all wisdom, and always present in eternal Now! All thoughts and things are within itself. It is **ALL** in **ALL**.

We are living in the day and time where **all things are possible**. There is no need to constantly think about what it is you do not have or you do not want. Time to throw this kind of thinking out! Take back your Mind as **Qatum (Melaninite) Quantum Beings!** Time to take on a **"NU"** or **"NUN"** type of thinking; we are all the children and descendants of a great Ancient Race of People. Throughout **Ourstory** which is called **History** the facts show how we built vast Kingdoms all through out what is called Africa today and in the Americas. Everything we are seeking from life we already have, this **Divine Mind** of ours is our birth right a gift from the Divine! The time is Now!

The Planes of Reality (Consciousness)

There are an Infinite number of Planes throughout existence, on this side of **Hydrogen** (**H1**) there are a total of **Nine**, for each of the *"Nine Energy Seats of Consciousness"*(**Chakras, Arushaat, Wheels of Life**) found in the African Physical make up known as **"Seats of Awareness"**. As each energy seat of conscious awareness opens up, you begin to perceive on higher **Quantum** levels. You perception begins to shift more and more to a more positive life frequency.

Question: What are Planes?

Answer: The most common mistake is imagining planes as layers of Strata or fine matter lying one above the other in Space. Planes are distinct and intricate **Moods Of Vibration**, interwoven into one another so uniquely that one is not able to determine where one plane begins and the other ends. Nothing in creation rests, Everything you Hear, Taste or Feel **Vibrates** .This is a well known fact now known to Modern day **Quantum Physicists**, and was always known to our Ancient **TaMa-Rean (Egyptian) Ancestors**, as the **Grand Hierophant Tehuti** taught us in his doctrine of **the Kybalion** also known in this day and time as **the Sacred Wisdom of Tehuti**. The natural **mood of vibration** ranges from very high and fast to very low and slow , There Are **Nine Planes** Or **what the religious Theologians are calling Heavens,** The Lowest And The Slowest Is The **Material Or Physical Plane,** While The Fastest And Highest Is The Plane Where One Is In Total Union And Bliss the infinite Cosmos.

Question: When these Religious Theologians where speaking of Heavens, did they really mean Planes?

Answer: Yes, when they were speaking of the Heavens, because of their limited perception of the **laws of Nature** and **Quantum Physics**, what they where referring to was the **Mental Plane** and **all the Planes above the Mental.**

Question: What do you mean the Mental; can you describe all of the Nine Planes for me?

Answer: Sure, the Nine Planes and their detailed explanations are listed as follows:

1. The Material Plane or Physical Plane:

The Material Plane of the Physical Plane is the plane that Human Beings dwell on. The Material Plane is governed by the law of Gravitation which is Desire. The Material Plane is partly made up of Solids, Liquids, and Gases. If you feel in your Mind that nothing else exist outside of the Material Plane you are wrong, there is more to life and this vast Universe then you realize. There is still more to this Plane then most People realize. The Western Science world today limit themselves things that only substances residing on the Material Plane is Solid, Liquid and Gas when in reality there is more.

Not only does there exist Solid, Liquid, and Gases as we have been taught but also four higher physical Etheric Substances which are:

1. *Ultra-gaseous Matter*
2. *Life Force Ether*
3. *Light Ether*
4. *Mental Ether*

2. The Plane of Force (Forces):

The **Plane of Force** (Forces Plane) is perfectly woven into all **Matter** on the **Material Plane**. The Basis or Mother of **Matter** is **"Nature"** and Nature is resulted from the **Plane of Force**. The Plane of Force is governed by a very **"Positive"** Force called **Attraction (A Law in Nature)**. The **Plane of Force** and the **Material Plane** (Physical Plane) work together in perfect **Harmony**. If this were not true, the *"Cycles of Nature"* would not be completed so perfectly and faithfully each year. The **Plane of Force**, also called the **"Energy Plane"** consists of ordinary forms of energy like **Heat, Light, Attraction, Magnetism, Electricity**, etc. As well as the forms of **Energy** which are finer, that gives Human Beings and Nature vitality and the ability to grow and reproduce. This Vital Life Force, which sets life on the Physical into motion and is responsible for growth and reproduction, is called *"Nafas"* in **Ashuric / Syric Arabic**, also known as *"Prana"* to **Hindus**, and in Ancient **TaMa-Re (Egypt)** it is named *"Sekhem"*, meaning **Power** and In **Hebrew** the word *Nafesh* mean *"Breath"*. Because the **Plane of Force** consists of **Etheric** duplicates of things on the **Material Plane (*Physical Plane*)** or counter parts, the **Plane of Force** and **Material plane (*Physical Plane*)** work together in perfect *"Harmony"*. The things you are seeking to Manipulate on the **Material**

Plane (*Physical Plane*) contain the same **Ethereal Substance** or **Material** as that, is on **The Plane of Force.** The "*vehicle*" of this **Plane of Force** in Humans is the **Subconscious Mind**.

The **Plane of Force** is all around us, and is as much a part of our makeup as the **Material Plane** (Physical Plane). There are Forces on the **Plane of Force** many have yet to uncover and comprehend; these are the Forces that control the Forces that result in what we call "*Nature*". The Lower Forces consists of ordinary forms of Energy like **Heat, Light, Attraction, Magnetism** and **Electricity**, etc... But the **Higher Forces** are those forms of Energy which are finer, they give Humans and Nature vitality and the ability to grow and reproduce.

The Basis or Mother of Matter is Nature and **Nature** is a result from the **Plane of Force.** The **Plane of Force** which is energy in all living things, or inside all living things. The **Plane of Force** access point is through the "**Subconscious Levels of Awareness.**" Your **Mind** feeds off the **Mental Reservoir (Universal Consciousness or Super Consciousness)** of Outellect. **Mind** is the individual part of you that feeds off the **Mental Reservoir (Universal Consciousness or Super**

Consciousness), the thought patterns of the **Soul** or **Spirit**.

3. The Spiritual Plane:

The **Plane** of **Emotional Energy** or **Energy in Motion**, also coming from the word Spirit meaning "**Breath, life**" as in **Life forces Vigor**, and **Ether**, or **Etheric State**, this is the **Plane** that also resides **within you** and your **Emotional Energy** is the **governor of this Plane**. **Spirit Beings** as some may call it, or **Etheric Energies Vibrate** at different **Density Levels**, and **Vibrational Rates**. In order to know what type of **Spirit Forces** you are attracting into your life, just check you're **Emotions**. Spirit or Etheric Forces that vibrate at lower Vibrations, tend to be attracted and feed off very depressed and repressed feelings. Feelings like Hate, Lust, Greed, Envy, and Jealousy. On the Spiritual Plane Human Beings do have a set of Sense just like the your 5 physical senses with this one difference, your Spiritual senses are capable of remembering your Physical experiences as well as your Spiritual experiences on higher Abodes. It is on this Plane that you are able to direct your Mind Power to Contact Ancestors who have passed from a Physical State on to the Etheric Worlds. Through the Powers of your Sub-Mental also known as you're "*Subconscious Minds*" you are in connection with your passed ancestors and can call upon their Etheric Powers, to aid and guide you. Call on them, they await your call only when you learn to Mental align with the "**Positive**" 9ether forces in this day and time.

4. The Mental Plane

This is the **Plane** within **Human Beings** that you are calling your "**MIND**" as in the word "**Mine**" and "**MIND**" sound phonetically the same, for it is your **Mind** that feeds off what is called the **Universal Mind**, or **Universal Consciousness**, **Mental Reservoir** of **Outellect** and even known as the **Super-Consciousness**. Like the old saying goes "**There is Nothing Nu (NEW) under the Sun (PAA RE)**." So this is where your thoughts of inspiration comes from, and depending on a person's **DNA** quality and Genetic makeup determines the Quality of thoughts a person will have and what part of the **Mental Reservoir** of **Outellect** they will be channeling from.

We as **Qatum (Melaninite) Beings**, have access to ALL, infinite and boundless. We have the ability to **channel** all of our **Ancestral Forces**, and bring them forth in this day and time, thus the rebirth of the GODDESS, and GODS, it is only up to the individual to realize your **infinite potential.** This is one of the many reasons of even Scribing a book like "**What is**

176

Qatum (Melanin) Physics?" To get your minds ready and introduce the Infinite Possibilities of who and what you are by **Nature** (Neter). Your Mind Power linked to the **MENTAL PLANE** is your **GOD power**, and your ruling power, now is the time to learn to use your **Mind Power** purposefully, or for a **Divine** and **Noble** purpose. Use your **Mind Power** to pull yourself and all others around you out of the rut, the Muck and mire on to Deity ship, use your **Mind Power** to **Visualize** the **Supreme Beings** you are here and right NOW!

5. The Plane of Truth or Divine Truth:

Where Truth begins to Manifest itself to you, were you begin to become Conscious and Aware of about the truths of life and it's many divisions, parts, and pieces. So the **Plane of Truth** within one self is what you are calling "**ME**", I exist, what I am, what I think, I am, what I want people to think I am, **My Persona** or **Personality** etc… that's the Truth, but then you move on to reality, "**what are you really?**" and this Answer only manifests to you and you alone, because you can't lie to yourself. Once you get to this point in your **SOUL'S** growth and Elevation then you move on to the **Plane of Reality!**

6. The Plane of Divine Reality

Where then you move on to the next **Conscious realization**, of what you thought was "**TRUE**" was accepted as truth at one time but not necessary a reality. An example of this, was that it was taught in old Europe that the world was flat and accepted as a truth at one time in Europe, but the reality was in **Ancient TaMa-Re (Egypt)**, we had pictures of the **Planet (PTAH-NUN) Earth** drawn as a Circle long before any other Cultures came into **Egypt** seeking to learn. So we did not accept this "**TRUTH**" as Facts which is confirmed "**REALITY**". It was also accepted as "**TRUTH**" amongst the Science world that **Atoms** was the **Lightest Element** when in reality they have now figured out that this is not true, the reality is **Atoms** are made up of smaller more minute particles called **Quarks**, then on to **Zeles**, then on to **Biaps** or **Mesons**. So yes at one time you "**consciously**" dealt with truths until your **Soul** begin to Elevate and you started seeking facts and realities about who and what you are, and your soul purpose during this incarnation as a **Qatum (Melaninite) Whole Light Being.** Now you have moved up to the Conscious realization of **GODLYNESS!** You are now at the Bosom of **GODSHIP**, taking control of your Life! The place where no lies come in because everything becomes a reality to you.

7. The Bosom of the Most High:

GODDESS SHIP, GOD SHIP, DEITY, you become your own Best Master, and Judge! Now you put on the **Mask of GOD** and you turn around as a Evolved Soul and you look back, and what do you see? You see "less" then yourself, because you are now at the seat of **GOD**, meaning you are at a conscious reality, that there are People on the **Planet Earth** (**PTAH-NUN**), or **SOULS** and **Spirits** less evolved or elevated then yourself.

8. The Plane of Change:

Once you have Elevated your Conscious realization to the point where you now know the reality of who and what you are, and you are now wearing the **Mask of God**, as a **"Responsible"** Human being, meaning having the ability to respond to the needs of others, then another realization comes into your conscious perception and reality. The **"Reality of Change"**, that change is the only real constant thing in the **Multiverse.** That what you where perceiving as death, and life, and growth is really change. So the reality of what they are calling infinity symbolized as the number 8 is really **"CHANGE"**.

9. The Plane of Completion:

The plane of perfect cycle, completion, where all realities meet. You are now a total complete being, knowing who and what you are, and when **divine purpose** manifest itself to you as "**Naught**". That which you "**NAUGHT**" to do, and that which you "**AUGHT**" to do. You move fully by the will of **Nature from within**, you can do nothing for or against yourself. **Aught** = Is, **Naught** = Aren't. Once you turn with the **Mask of God** on in your life and look back you begin to see the Illusion of what People, Places and Things "Aren't".

The Other side of H1 (Hydrogen)

10.The Plane of Aught:

Anti matter, the **Plane of Non-existence**, because once you complete the **perfect cycle**, the Cycle starts all over again but in reverse, so now your **Conscious Awareness**, turns to relating to persons, places and things of how you can improve upon that Cycle in your life from which you came from. This is when Reality, starts when you have not got passed "**Naught**" and the illusions of the World, and the way you consciously perceived persons, places and things. On the side of **Matter** there is **Anti-Matter**, on the

Side of **GOOD SPIRIT**, there is **EVIL SPIRIT**. A PUSHING AND PULLING, an Ebb and Flow, and a Expansion and Contraction.

11.The Plane of Relativity:

The realization of how persons, places, and things the seen and unseen are all inter-related by necessity, meaning they have a purpose, or it would not be so, the Balancement of ALL.

Question: Wow, this is a lot of Information; you keep saying these planes are in Us, How?

Answer: Great question and we knew you were going to ask that. Well when you begin to do a study of your **Anatomy** as a **Qatum (Melaninite) Being**, you will find what the Ancients are calling "**Energy Seats**", **Chakras, Arushaat**, or **Wheels of Life**, is another way of what Modern day Scientist call the "**Endocrine System**". Your **Endocrine System** comprises of **Glands** that are **intune** with Each **Plane of Existence** and have an effect on your **Consciousness** or **Mental State of Being.**

Depending on which Glands in your body are most active and vibrating the fastest, determines which Plane of Existence or Consciousness you are intune with, thus your perception on life.

Question: What are Glands?

Answer: A **gland** is a group of cells that produces and **secretes**, or gives off, **chemicals** known as "**Hormones**". Here is a list of the seven major **Chakras** most are familiar with and have been exposed to in this day and time.

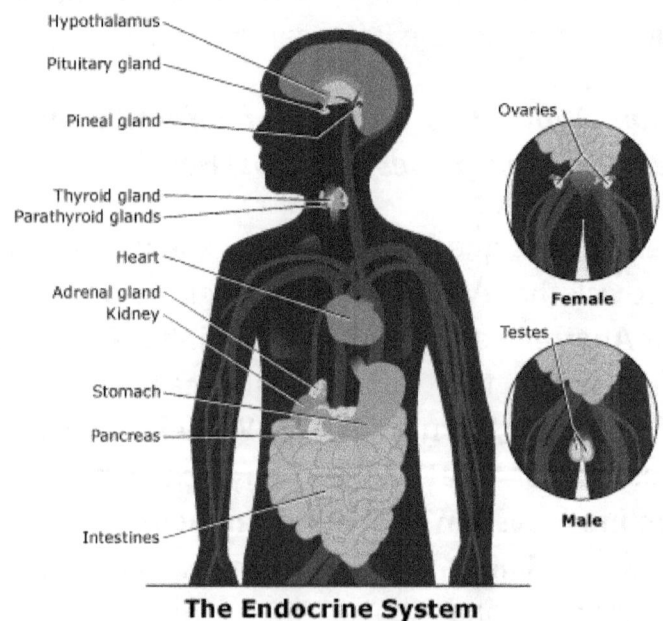

The Endocrine System

1. **Root Seat**: Superimposes the **Prostate area** in Males, and the **Uterus in the Female**.

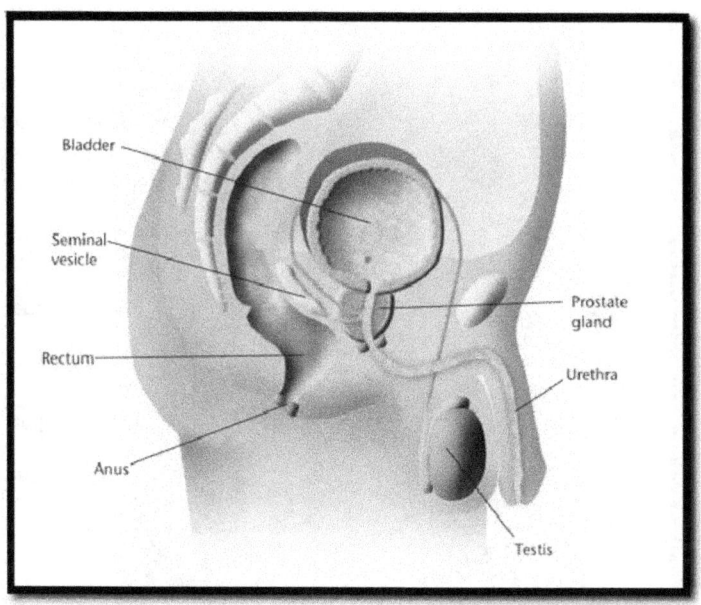

2. **Splenic Seat**: Superimposes the **Navel** and Extends to imbed some of its roots in the **Spleen**.

3. **Solar Plexus Seat**: Superimposes the **Solar Plexus** area and the **Great Lobe of the Liver**.

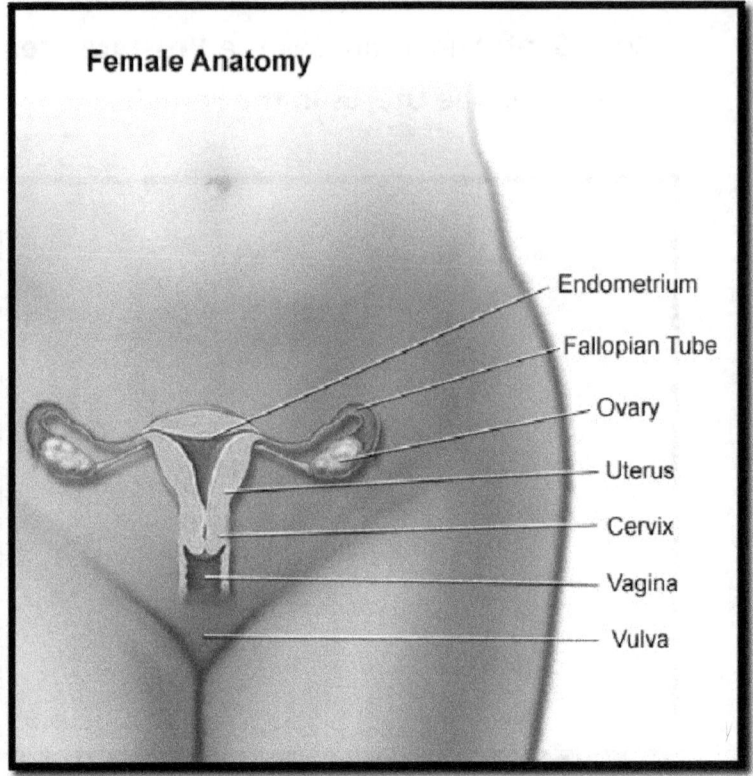

Female Anatomy

- Endometrium
- Fallopian Tube
- Ovary
- Uterus
- Cervix
- Vagina
- Vulva

4. **Heart Seat**: Superimposes the **Thymus Gland** between the shoulder Blades and the **Right Auricle of the Pulse Point of the Heart**

5. **Throat Seat**: Superimposes the **Thyroid Gland** in the Throat, and extends to embrace the Medulla Oblongata.

6. **Brow Seat**: Superimposes the **Pituitary Gland.**

7. **Crown Seat**: Superimposes the **Pineal Gland. Radiates Outward to cover the entire top of the Head.**

*Note in our Ancient TaMa-Rean System we knew about 2 more Energy seats, thus a total of NINE (9) that makes you a Whole Complete being.

1. The Carnal – Also known as the Perineum

2. The Nasal – The Nose, tip of your Personal Pyramid that ignites the 3rd eye or Pineal Gland.

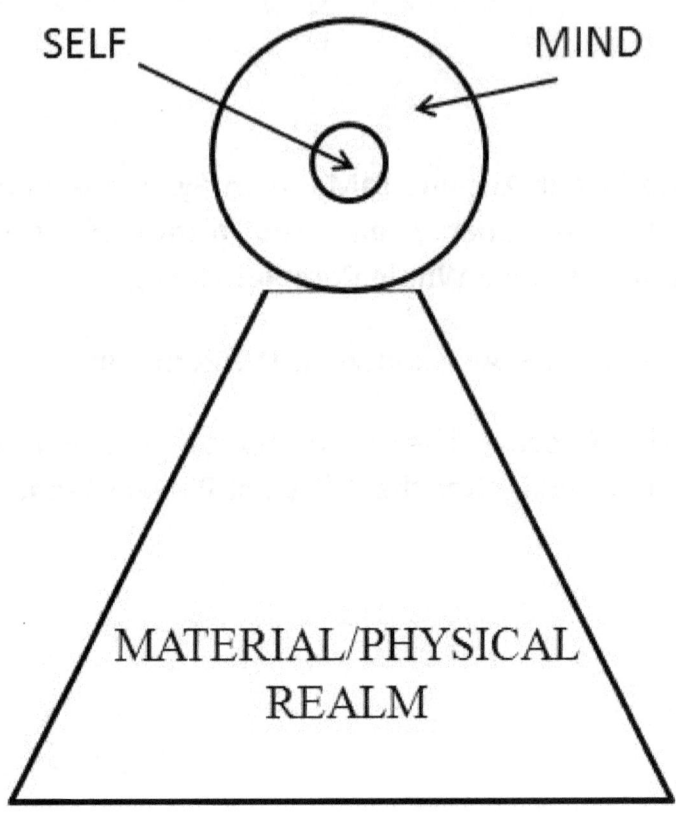

SELF   MIND

MATERIAL/PHYSICAL REALM

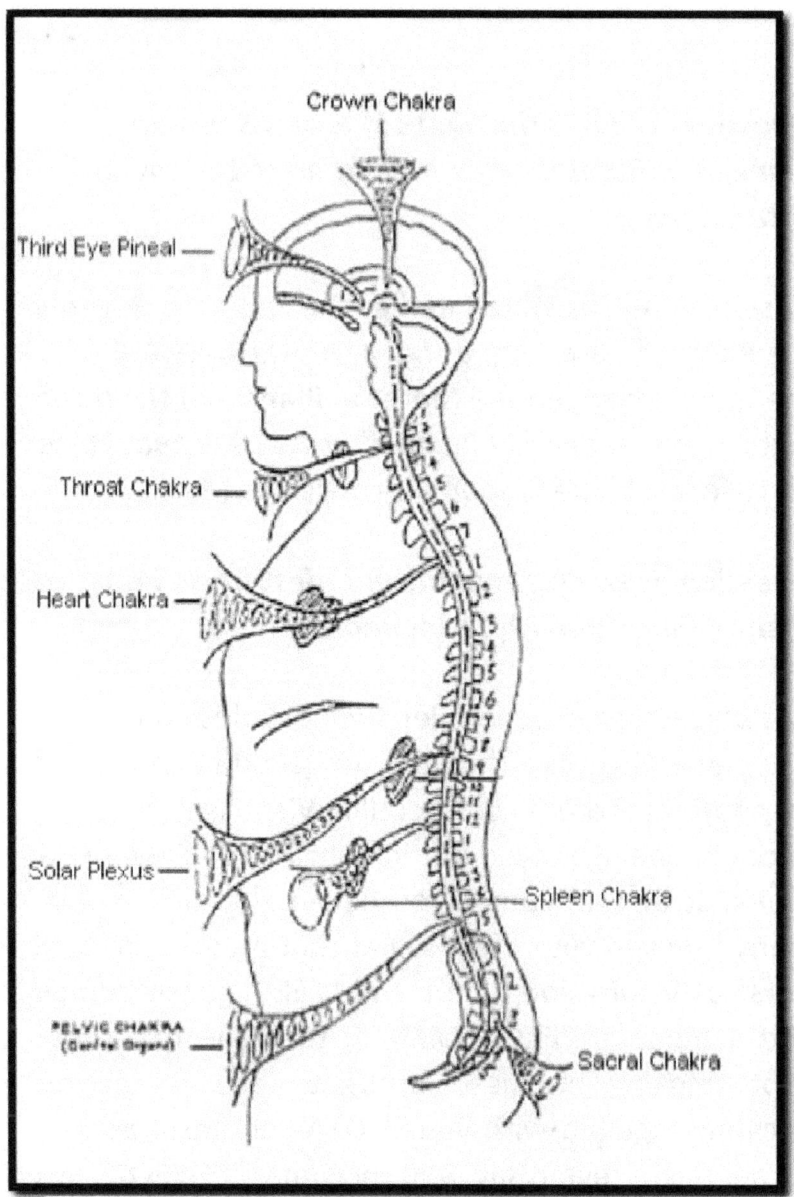

The Energy Seats

Question: What is the Plane of Force, Spiritual Plane, and Mental Plane in relation to the Law of Attraction?

Answer: Well, as stated earlier the Law that Governs the **Plane of Force** is the **"*LAW OF ATTRACTION*"** and the bridge between the **Material Plane** and the three higher planes, be they **Plane of Force**, **Spiritual Plane** and **Mental Plane** is your **Subconscious Mind.**

Question: How do Thoughts manifest from the Mental Plane into Material Reality?

Answer: Great question, let's view it this way. Have you ever had a **Thought** or sudden burst of **Inspiration**, where you could **Visually** see your **Thoughts** being played out in your **Mind,** like more a **movie screen**, of course you have right, one time or many times in your life. Well at that moment of that **Burst of Inspiration**, you could "**FEEL**" goose bumps and a burst of "**EXCITEMENT**", then the more you kept thinking of these **thoughts**, the more you became engulfed with your "**IDEA**" or "**Inspiration**", Some might even say you became "**POSSED**". So finally as you kept thinking of the this inspiring idea, you begin to think of a **plan of action** on how to

Materialize this Idea. This plan of Action is what is termed **"ATTRACTION"** you begin to move on the energy or react on the energy that was inspiring you into **Action**, you went from **Static Energy** to **Kinetic Energy** which is **Energy in Motion** or **Emotional Energy**. As you stayed focus on your Idea, and kept the high **Emotional drive,** you begin to see little things manifest in your life that let you know you are on the right track. These are the little **"magnetic magical particles"** manifesting in different ways in your life from the **Plane of Forces**. You might have seen your Idea manifest or **Materialize** in many ways, like **intuitive flashes** in your **Mind**, as People to call, Places to go or visit or even things to go buy that would assist or aid you in Manifesting this Idea and inspiration you had into **Solid, Concrete**, and **definite Reality**. Then as you listened and followed you finally seen the Manifestation of your **Thoughts** from the **Mental Plane** come forth in what seem like nowhere, which really means **NOW-HERE!**

So let's say this another way to make sure that this is **"CRYSTAL CLEAR"** in your **MIND'S EYE** as to how your **Thoughts** are **Manifested** into what is being termed **"CONRETE REALITY"**, thus as the old saying goes **"Thoughts become Things"**.

Manifestation of Thoughts
from the UNREAL TO REAL

1. First you have a Burst of inspiration, **IDEA** or **Thought = Mental Plane.**
2. Then you "**FEEL**" the goose bumps or **Emotions** attached to this **Inspiration = Spiritual Plane (Plane of Emotional Energy)**
3. Then you take Actions , meaning your Emotions have now become active or "**IN MOTION**" thus **Energy in motion**, and as you **Move** or **ARE IN MOTION**, you create **FORCE**, for **Mass=Force.** As movement occurs it cause **unseen Atomic Particles** to **Magnetically** "**ATTRACT**" to your **Holographic IMAGE** or **IMAGES**, coming from the **Black SPACE** of your **Mind (DARKNESS). Magnetic Attraction** and **FORCE = Plane of Force (LAW OF ATTRACTION).**
4. Then you being to see the manifestation in the three Solid visible realities as **Persons** you might meet to assist you into bring this **IDEA INTO SOLID REALITY, Places** that you might go in order to Bring this **IDEA** into **Solid Reality**, or things you might need to acquire in order to bring this Thought, **IDEA**, or **Inspiration** into **Solid Reality. THUS MANIFESTATION OF THOUGHTS!**

Question: So how important are your Ideas?

Answer: Our Ideas are very important and need to be taken seriously. Our **Ideas** and **Thoughts** are **living entities**; they are alive in a more finer place within existence. Like we taught in **Volume 1 "An Introduction to "NU" Thought"**, our thoughts and ideas are **"Spiritual"** in **NATURE**; they consist of a finer substance called **"Ether"**.

What you see in your **Mind's Eye**, is a real living **"Entity"** known as **"Thought Forms"**, which is why **corporations, media, television** and **magazines** spend **billions upon billions of dollars** making sure their images 1st seep into your **Mind**, through **advertisement** (adverse –tise, ment = **meant to entice adverse forces within**) once these false images that are not truly aligned with your life's **purpose** and **Souls destiny** are in your **Mind** and seep into your **Subconscious**, they are then being played out. A lot of times people are not even **Conscious (aware)** of the things they are doing or why they "**Desire**" certain **foods, brands of clothing, music**, etc... This is just an example of how these same **mental formulas** have been used in a **negative way**. In this new **Millennium** known as the "**Solar Cycle of RE**" or simply **Sun Cycle,** People (**Human Beings**) will become more **Conscious** (**aware**) of the

Quality of **Thoughts** they are having and the kind of **Information** we allow to *"enter into our Minds".* *(Refer to Solar Cycle of RE – Master Key Vol. 2)*

Now we living in a more **"positive"** **Cycle of Nature**, where Father Earth in ancient **TaMa-Re** **(Egypt)** known as **"Geb"** which means **"Time"** because times does not begin until you are born to this Earthly realm or your **Conscious Mind** does not perceive **Time** until you are subjected by the **Laws of Physical Reality**. Many have been taught that we lived in tune with **Mother Nature** and **Father Time** they called the **Universe**. The time is now to view the **Universe** as your **Mom**, **Mother**, and **Matter**, **Space** in ancient **TaMa-Re** **(Egypt)** was seen as the **Black Womb** of the **Universe** known as **Mother "NUT"**, and you as a **child of the Universe** can **impregnate** your **Positive Thoughts** into her Womb and watch your **thoughts** manifest into **Physical Reality**. This is the **Great Day and Time** we are now in! Be part of the **Upward Swing!**

Question: Can you explain to me further about the Law of Attraction and how it relates to us Qatum (Melaninite) People?

Answer: Well, According to the Merriam Webster's Collegiate Dictionary the word "**ATTRACTION**" is defined as: **To pull to or draw toward oneself or itself.**

You as **Qatum (Melaninite) People** are very "**ATTRACTIVE**" because of your **Melanin** which absorbs the full light **Electro-Magnetic Spectrum**, So by **Nature** we have always been able to pull and draw things to us, meaning pulling or the word we like to use now "**CHANNEL**" awe inspiring **Outformation** and **Revelations** that would have a **Positive,** lasting effect on the **World** and **future generations.** So as we begin to learn further about the "**Law of Attraction**" we learn that **the Law** teaches us that "**likes Attract likes**" maintains in the **Thought – World**, and that one "**Attracts**" to him or herself the **Thoughts** of others which correspond in kind with those held by him or her.

Have you ever wondered why certain people of certain professions gather together, like Actors hang out with Actors, or Musicians hang out with Musicians etc... The old saying rains true "**birds in a feather, flock together**", like Minds "**Attract**" to each other. Now let's take time to examine the "**Attractive Power of Thought**."

As we mentioned earlier around every person is an **Electro-Magnetic Field** of Energy that is influenced by **Thought Waves** coming from either the individual, other people or ultimately your environment. Every thought that we think starts in motion **Thought – Waves**, or **Vibrations (Vibes)** which travel along with greater or lesser speed and intensity varying with the "**force**" of the original **Thought** and which affect, more or less People far removed from the persons sending forth the **Thought**. We as Humans are constantly sending forth **Thoughts** and are constantly receiving **Thought- Waves** from others. A Person who "**hates**" will "**Attract**" to themselves all the hateful and malicious **Thought –Waves** within a large radius, and these added **Thoughts** act as fuel to the fire of his or her base feelings, and render him or her more hateful and hating than ever, which is why as **Qatum (Melaninite) Beings** it is very important in this day and time to **Mind your Thoughts**, **Feels** and **actions** because we are the most **Magnetic** and **Electric** on the **Planet Earth (PTAH-NUN)**.

Introduction to the Subconscious Plane of Thinking

Question: You mentioned the word "Subconscious Mind" several times, can you tell exactly what it is?

Answer: Sure, we where leading into introducing you into the **Subconscious Mind** and its functions. What most People need to begin to realize in this day and time, is there are let's say three for now divisions of your **Mind** that are all linked to one great stream of Pure Consciousness known as the "**Mental Reservoir**", in Ancient Sanskrit Language this **Mental Reservoir** is known as the **ASKASHA**, or **AKASHIC** records, meaning in Sanskrit "**aether**" or "**ether**" in ancient **TaMa-Re (Egypt)** this "**Pure Stream of Consciousness**" is known as **PAUT "ALL"**.

When we look at the word **Sub-Conscious** we see two words "**Sub**" meaning **below** or **under**, and **Conscious meaning awareness**, So **Subconscious** in the English sense of the word means "**Below Awareness**". As you already know by now you are **QATUM (Melaninite)** and **Quantum Beings**, and there is a whole vast existence going on right "**underneath your own conscious perception and awareness**".

Question: Can you give more information on our Subconscious Minds and its role?

Answer: Sure this Section below goes into more in-depth information pertaining to the *"Infinite Power of our Subconscious Minds"*:

The Subconscious Mind

Sub = Below, **Conscious** = Awareness (**Below Awareness**)

Subconscious: The Mental activity just below the threshold of Conscious Awareness – The Seat of the Fourth Dimension (The Plane of Vibration)

Your **Subconscious Mind** is the seat of your **emotions** (**Energy in Motion**), in the **Physical body** the **Subconscious Mind** is termed the **Abdominal Brain**, located just **three inches above the navel** termed the *Solar Plexus (refer to the Book "The Solar Seat of RE")* and is the **Creative Mind**; when you think good, good will follow; when you think evil, evil will follow. This is the way your **Mind** works. The point to remember is once the **Subconscious Mind** accepts an idea, it begins to execute it. It is an interesting and subtle truth that the **Law of the Subconscious Mind** works for good and bad thoughts and ideas.

This law when applied in a **negative way** is the cause of failure, frustration, and unhappiness. However when your constant thoughts are harmonious and constructive, you experience perfect health, success, and prosperity. Peace of mind and a healthy body are inevitable when you begin to think and feel in the right way.

Whatever you claim mentally and feel as true, your **Subconscious Mind** will accept and bring forth into your experience. The only thing necessary for you to do is to get your **Subconscious Mind** to accept your idea, and the law of your own **Subconscious** will faithfully reproduce the idea impressed upon it. The law of your mind is this: **You will get a reaction or response from your Subconscious Mind according to the nature of the thought or idea you hold in your Conscious Mind.**

Think the things you would see manifested, see them, know them, and you can leave it to your **Subconscious Mind** to bring them into being. For thought is *creative energy*. It brings into being the things that you think!

(*Refer to the: Holographic Brain – 4th Dimensional Reality Master Key Vol. 18*)

The Subconscious Mind – Doorway to ALL

The **Solar Plexus** is the point where **life is born** – where the uncreated becomes create; the unorganized becomes organized; the unconscious becomes conscious; the invisible appears; that which is dimensionless becomes measurable. The **Solar Plexus** is the **Seat of the Subconscious.**

Until we overstand and take control of ourselves, every thought that passes through the **Mind** affects the action of the **Solar Plexus**, and the **Solar Plexus** is the **seat of the Subconscious Mind** called the **Abdominal Brain. Non- Resistant thought expands the Solar Plexus; Resistant thought contracts it.** We must learn to control the action of the **Solar Plexus** just as we learn to control the action of the fingers in learning to play the piano; by **thought** and careful exercise. Make up your Mind to keep your light shinning, your **Solar Center** expanded, no matter what happens or how you may feel. The **Solar Plexus**, or **Sun Center** or **your Central Sun**, is to the **Human Body** precisely what the **visible Sun** is to the **Solar System**. It is the Source of ALL life and Light. It is the Manufacturer of Life and Light. The **Solar Plexus** breaths in intelligent **will,** and breaths out **magnetism.** Remember breathing and thinking are one. **To think deep, Breath deep!**

Your **Intuition** is the voice of the **Subconscious Mind,** which is linked to your **SOLAR PLEXUS** that gut feeling (**vibration**) in the pit of your **Stomach,** which is linked to ALL residing on the **Plane of Force** and operating your **Psychic faculties.** (***Refer to the Solar Seat of Re Book***).

The Electrical Magnetic Power of Thought

Thoughts operate on the principle of **Electrical Energy. Electrical Energy** is of a special importance in **Energy** changes. Most forms of **Energy** can be changed into **Electrical Energy. Electrical Energy,** in turn can be changed into almost any other **form of Energy.**

All around you is **Energy − Electrical Energy,** exactly like that which makes up the solid objects you possess. The only difference is that the loose Energy around us is unaccounted for. It is still virgin gold − **untouched, undiscovered** and **unclaimed.** *You can think it into anything you wish −* **great health** *or* **sickness,** *into* **strength** *or* **weakness,** into success or failure. What will you form this Energy into in order to live your life's purpose? There is nothing good or bad but **Thinking** makes it so! The **Overstanding** of this **Law** within Nature will enable you to control every other **Law** that exists. In it is to be found the "*Universal Remedy*", for all ills, the satisfaction of all want, all needs and "**Desire**". It is *Creative Mind's* own provision your freedom. **Today is the start of the rest of your life**, this statement simply means you always have the power to start over, to remake or renew your life.

How do you see yourself today? Where are you going and what do you truly want to do with your life? What is your true life's destiny? You can learn through the powers of your **Subconscious Minds** to find the answers to these questions and many, many more. The universe is boundless, and infinite energy, this energy again is just waiting for us to tap into it. How will you use your **Mind Power (Thought Energy)** to shape this "**Electrical**" **Energy**?

Visualization Exercises, Auto Suggestions, The Power of Spoken Word on the Subconscious Mind

Question: As Qatum (Melaninite) Beings what effect does Visual images have on the Subconscious Mind and the Quantum World of Reality?

Answer: The **images** that you hold in your **Mind**, have a major effect on the **Quantum fields of reality**, because those **images** are **holographic projections**

stored by **the Subconscious Mind**, created from Various scenes from your everyday life and or **inner inspirations**, or **outer influences**. **Visual imagery** has a major effect on your over all **emotional being**, which seeps down the **Subconscious fields of awareness.**

Let's take a something very interesting also when you look at the word "**imagination**" you see the word "**Image**" (image-nation) in this word. So **imagination** is the **creative ability** of the **Mind** to form **visual mental pictures** or **images.** Another word for this process of seeing **mental images** is called "*Visualization*".

Now, there is a very real "**Law of Cause and Effect**" within Nature which makes the Dreams of the Dreamer come true, and this Law is the "**Law of Visualization**". Imagination pictures the object you desire, but the power of "**Vision**" idealizes it. Imagination gives you the picture, but Vision gives you the impulse to make the picture your own. Start today that you truly have the power to bring about your "*ultimate success*" just by the images you hold in your mind.

Make your mental images clear, and vividly see them in your Mind's Eye, then just relax and forget about the whole matter. Send these visual images over to the **Subconscious Mind**, by relaxing and letting go of the images you hold and watch them began to manifest into everyday reality. Visualization, my Brothers and Sisters holds the **Ankh** (☥) Key to being able to shape and mold your own destiny!

Creative "Visualization"

Thought externalizes itself. What we are depends entirely upon the images we hold before our *Mind's Eye*. Every time we think, we start a chain of causes which will create conditions (effects) similar to the thoughts which originated it. Every thought we hold in our **Consciousness** for any length of time becomes impressed upon our **Subconscious Mind** and creates a pattern which the **Mind** then weaves into our life or environment.

All power that we seek is from within and is therefore under our control, once we learn to discipline our Minds. When you are able to direct your thought processes, you can *"Consciously"* apply them to any condition, for all comes to us from the World outside of us from what we've already *"Visualized"* in the World within (our Minds). Do you desire more Money? Sit down quietly right now and begin to realize with your Mind's Eye, that *"**Money**"* is merely an Idea. That your Mind is possessed of unlimited ideas, that being part of **Universal Mind**, there is no such thing as *"limitation"* or lack. That somewhere, somehow, the ideas that shall bring you all the Money you need for any right purpose are available for you. That all you have to do is send the thought frequency to your **Subconscious Mind** to find the "**Supreme**" answer for you. Realize it, Know it, and your needs will be met.

"Whatever things you Desire, when you ask for it from your **Ancient Ancestral Forces** know that you will receive it. They are linked to you, linked to all of us through our **Right Minds** (**Subconscious Minds**). "**Ask and yea shall receive**". Know that as Qatum (Melaninite) People we can heal the World! We have done it before and now we will do it again. "It's over due time for all the "**Good Hearted People**" to come!" Use the gift of your **Mind, the power of** *"Creative Visualization"* where you create pictures and whole movies in your Mind of what you desire to manifest.

Practical Use of Creative Visualization

Exercise #1: Self-appreciation Meditation

Here is a very helpful, practical and useful meditation and creative visualization exercise you can do to improve self-esteem and increase your capacity to handle the divine love and energy that the Universe (Multiverse) is abundantly ready and eager to flow in your direction:

First find a quiet place either in your Home, office, in the Park or even a Place that is Sacred to you. Sit back in a relaxing chair, on the grass, or you might even lie down on your bed, or futon. Just make sure you sure you are comfortable, relaxed and free from distractions. You might dim the lights if you like, or even light a candle, or even burn your favorite scented oil or incense. Just allow inner guide set the stage for you, it truly knows best. Now once you are all set and feel relaxed let's begin shall we!

"Close your eyes and start to breath slowly, inhaling through your nose and exhaling through your mouth, slowly and steady, just allow your inner guidance to guide you, it knows best. Now as you been to breath slowly and feel relaxed being to imagine yourself in some everyday situation, and picture someone (maybe someone you know, or a complete stranger) looking at you with great love and admiration and telling you something they really like about you. Now picture a few more People coming up and agreeing that you are a very wonderful person. (If you begin to feel embarrassed it's ok, just stick with your visualization and it will subside.) Imagine more and more People arriving and gazing at you with tremendous love and respect in their eyes. Picture yourself in a parade or on stage, with multitude of powerful cheering, applauding people, all loving and appreciating you. Hear their applause ringing in your ears. Stand up and take a bow, and thank them for their support and appreciation."

After you finish this **Creative Visualization** Exercises, just sit back and relax, and calmly come out of your meditation, then you are finished.

*Note: Practices Exercises #1 from time to as you feel fit, learn to listen to your inner voice, your "**inner guidance system**", your "**intuition**" which is the voice of your **Subconscious Mind**, which is linked to ALL, it knows best, and when you need to do this exercise. Know that repetition is the "**key**" to success and that it is best to practice this exercise and future **Practical Visualization** exercises upon retiring to lay down to go to sleep, and upon waking up from a restful sleep. These are the times the "**doorways**" to your **Subconscious Minds** are most **open** and **activate**. Remember whatever you impress upon your **Subconscious Minds** becomes part of your Outer **REALITY**, and everyday living.

How to Release Tension and Stress from your Everyday Life!

Practical Visualization Exercise #2:

In this day and time with all the many stresses of day to day living, we tended to carry so much tension and emotional stress in our being that we are even consciously aware. These tensions and Emotional stress ultimately have a "negative effect" on our Melanin and our whole physiological makeup. With this **Practical Visualization Exercise** we can begin to learn to relieve built up tension and stress, for our ultimate Health and Well being. As you know these **Practical Exercises** go a long way in restoring balance not just in areas of the **Physical Body**, but also in the **MIND (Mental)**, **Spirit**, and Soul (**Emotional**) as well. Remember we are **Qatum (Melaninite)** and **Quantum Beings** and we need a system of restoration that deals with all four aspects of our being, **Mind (KHU)**, **Body (KHAT)**, **Spirit (KAA)** and **Soul (BAA)**. Now let's begin shall we!

"Just before laying to rest, turn off all lights, and sounds in your room or dwelling, and begin your breathing exercises, relaxing and calming breathing through your nose and out your mouth, in a slow but rhythmic breathing pattern. Follow your inner voice, it knows best, just relax and be calm. Now as you begin to close your eyes focus your attention on the "darkness or blackness of infinite space" that you see in your Mind, try to determine where the darkness begins and where it ends, for a split second you might feel you are traveling on a long journey to nowhere, and remember you are because the word nowhere really means "NOW-HERE" so realize you are in the NOW. As you begin to see the black infinite space in your mind turn your attention to the soles of your feet, feel a sensation in the soles of your feet and if you feel any built up stress or tension see it in your mind pulling away from your body.

As you feel tension being pulled away from your feet, breathe out slowly and then just relax, and as you breathe in visualize fresh new energy coming into your feet revitalizing your feet. As this energy travels into your feet, feel the energy traveling slowly up to your calves, then your knees, releasing all tension as it travels upward. Keep visualizing and feeling this Healing loving energy traveling up your knees to your reproductive organs, then up to your Navel (belly button) on up to the Solar Plexus region of your abdomen which sits three inches above your Navel.

As you feel a surge of increased energy and vitality, feel this energy travel up to your Heart as this Energy hits your Heart and Heart Chakra or Arusha or Energy seat, feel this loving energy opening your heart with Pure Green Divine Loving Energy (it's ok if you feel overwhelmed and you want to cry, this is normal this is Divine Love from the Universe –Multiverse). Now feel this Energy leaving your Heart area, and drifting off to infinite space. As you Visualize this Green Light drifting off into what seems like infinite and boundless space, then imagine and visualize a new energy coming or rushing towards you. This is the healing and loving force of your Ancestors being sent forth to you and towards you do not be alarmed, this Positive healing Green Light Force is here to heal you, remember Green is a color of healing, growth, vegetation and vitality. As you feel this New Green Light Healing force entering your Heart visualize this loving Divine love Energy, traveling up to your Throat area, opening all lines of communication for you and removing any blocks, now feel the Green Light Energy moving to the Great INNER EYE known as your "PINEAL" SEAT, 3RD EYE, OR PINEAL GLAND.

Feel this loving Green Light Energy, clearing any tension that sits in the middle of your head, as you feel this Green Light Energy, moving around in a Circular motion, feel it breaking up any unwanted Mental Patterns or unwanted Negative Thoughts. Feel a renewed Mind, and a fresh new start, after feeling this envision this PURE GREEN LIGHT HEALING

ENERGY, traveling back down towards your stomach and descending down into the Solar Plexus region where you can store it for use at any time!

You have just been re-introduced to the healing power of **Creative Thought.**

The Power of Spoken word on the

"Subconscious Mind"

Question: How do we speak to our Subconscious Minds?

Answer: The best way to speak to the **Subconscious** part of your being is by way of what is known as **"Auto Suggestions."** **Auto-Suggestions** is a term which applies to all suggestions and all self – administered stimuli which reach one's mind through your five senses i.e. **Hearing, Tasting, Smelling, Seeing,** and **Touch** which all come out to be One sense and that is **"Perception".** Another way of saying **"Auto-Suggestions"** is **"Self-Suggestion or Affirmation."** It is the agency of communication between that part of the **Mind** where **Conscious Thought** takes place, and that which serves as the seat of action for the **Subconscious Mind.**

Through the dominating thoughts which one permits to remain in his or her **conscious mind,** whether these thought be **negative** or **positive** does no matter, **the Principle of Auto-Suggestion** voluntarily reaches the **Subconscious Mind** and influences it with these thoughts. No Thought, whether it be **positive** or **negative** can enter the **Subconscious Mind** without the aid of the **Principle of Auto-Suggestion**, with the exception of thoughts picked up from **the Ethers.**

The **Power of Spoken** word upon your **Mind** is real, and the repetition of these words, be they **positive** or **negative** is also very real. Now just think for a minute, living in a society full of negative press, negative news, negative music etc... well you get the point for now!

Quantum Affirmation
for instant Life changing results!

Question: What is Affirmation?

Answer: According to the Merriam Webster's online Dictionary it means: Function: noun
Date: 15th century 1 a: the act of affirming b : something affirmed : a positive assertion

THE ACT OF AFFIRMING YOUR OWN EXISTENCE -
From your Conscious Mind to SUBCONSCIOUS!

So here is a life changing **Quantum Affirmation** - say it over and over again while falling asleep and in the middle of the night (when you wake up to use the bathroom). And watch you begin to affect the **QUANTUM FIELD** of not only your life but EVERYTHING, PERSON, AND PLACE AROUND YOU. Here it is: SAY...

I AM "IN" THE "LOVE" OF "ALL" AND
"ALL" LOVE IS "IN" ME
I AM PART OF "ALL" AND "ALL" IS PART OF ME
I AM "ONE" WITH "ALL" AND
"ALL" IS "ONE" WITH ME
I CAN BE "ALL" THAT I "WISH" IN "ALL"
AS LONG AS MY "WISH" IS TO STAY IN "ALL"
I CAN SUCCEED AS PART OF "ALL" AND FAIL AS AN INDIVIDUAL
I-AM-TRULY-NEVER-ALONE!
ALL IS... I AM
ALL CAN ... I CAN
ALL DOES ... I DO!
Say this **Quantum Affirmation** over and over again each night for about one week, just try it and watch seemingly **MIRACLES** "MIR-RA-KA'S" HAPPEN!

The Quantum Physics of the Pineal Gland

The Pineal Gland which is the Physical aspect of what is termed the third (3rd) Eye processes Light, from the Full Light Spectrum known as the Electro-Magnetic Spectrum from the SUN (PAA RE). This small Neuro-Endocrine Gland, the size of a Pea and the Shape of a Pine-Cone, has baffled modern science for decades. Te Pineal Glands creative functions have led Ancients to refer to it as the creative "Gene of Isis or Aset" (Genesis meaning Gene-ISIS).

ASET also known as ISIS

The **Pineal Gland's** partly **"crystallized"** Nature and ability to absorb, store and transmit Light store Energy and transmit Light Energy suggests it is the body's inner crystal. Science has recently discovered back in the year 2002 that the **Pineal Gland** is covered with **crystalline structures** called *"Calcite Micro-Crystals"* that runs of the science of *"Piezoelectricity"*. The word *piezoelectricity* means electricity resulting from pressure. It is derived from the Greek *piezo* or *piezein* (πιέζειν), which means **to squeeze** or **press**, and *electric* or *electron* (ἤλεκτρον), which stands for amber – **an ancient source of electric charge.**

*Note: notice the Light what part of the Light spectrum they Europeans definition acknowledges, the "amber light". Europeans, pull Pineal Gland pulls from a lesser Light Spectrum and Vibration, then you "QATUM (MELANINITE) Beings, But the fact still remains your Pineal Gland is a Crystal the 'BLACK CRYSTAL" which now you know why you have Telepathic Powers etc...

Now the Pineal Gland works closely with the Pituitary Gland. That's because these two Energy Seats (Chakras, Arushaat) are linked. The Pituitary is known as the "MASTER GLAND" because it has a stimulating effect on various bodily systems. The **Pineal Gland** we could usefully call the "**Mother Gland**" because it appears to calm the Body's Systems. The **Pineal Gland** releases **two Neuro-Hormones**, "Serotonin" and "**Melatonin**". **Serotonin** is a **Neuro Chemical** which is dominant during the day and helps keep us active and awake, while Melatonin encourages rest and is Dominant at Night. Together they regulate the **Body's circadian Rhythm,** which is **Biological Cycles** responsible for maintaining balance in relation to Light and Darkness.

When we stay in Darkness for several days, Serotonin and Melatonin Levels continue to alternate. If however we are exposed to constant Light or we stay awake for several days, the Rhythm is disrupted. Melatonin is associated with the release of Innate Hallucinogens, and research shows that High levels of Melatonin are present during altered "States of Consciousness" and Psychic visions.

Neuro Transmitters and "BRAIN CHECMICALS: for Heighten Spiritual Awareness:

Tryptophan(5HTP)-Serotonin-Melatonin-Pinoline-5-MeO-DMT-DMT

*Note: Melatonin is also active in the production of Melanin.

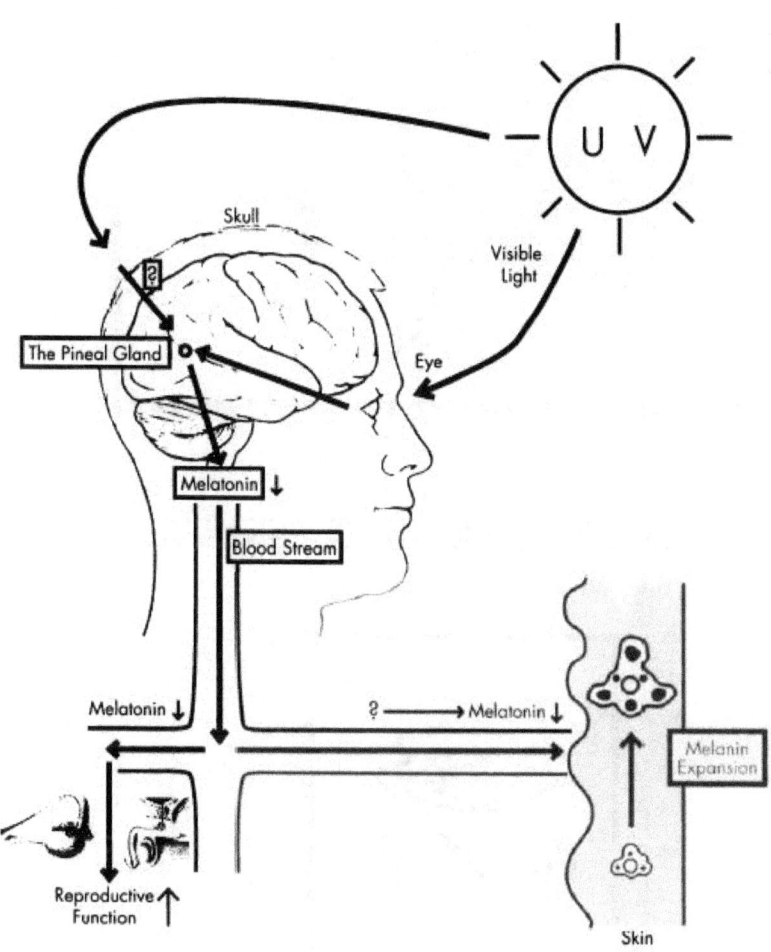

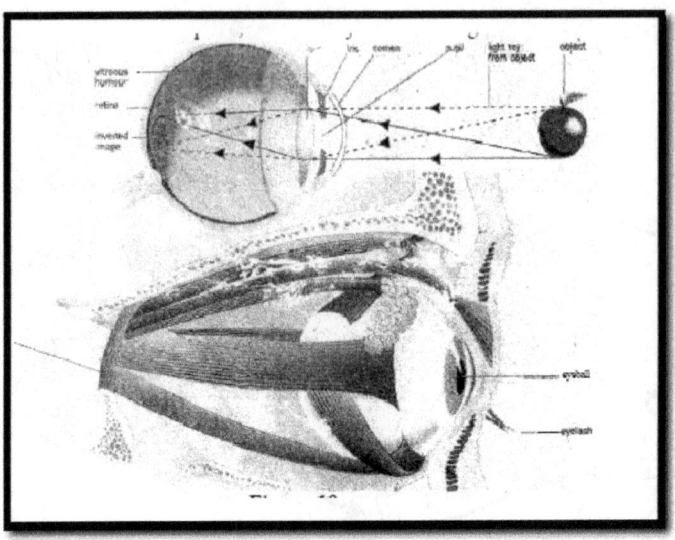

THE IRIS (Pupil) -THE "BLACK HOLE"

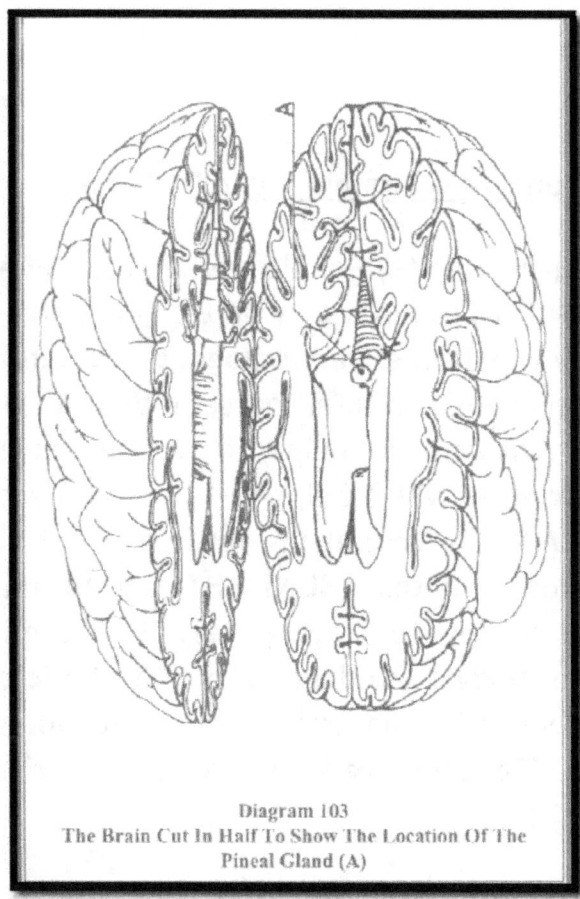

Diagram 103
The Brain Cut In Half To Show The Location Of The
Pineal Gland (A)

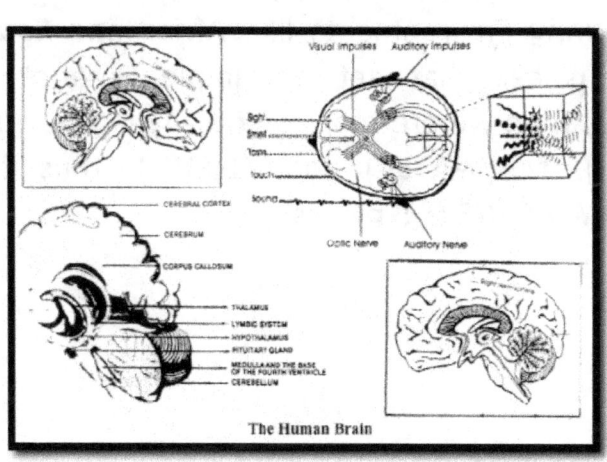

The Human Brain

Sunlight – Vitamin D and Melanin

As Qatum (**Melaninite**) People we need the **POWER of the SUN (PAA RE)**, not only do we get charged by the **SUN** by way of the **Electromagnetic Spectrum** of **Light**, but also it is a Natural source of **Vitamin D** and **Serotonin**. **Vitamin D** helps to strengthen your **immune sickness** which helps to fight off infectious diseases. **PAA RE (THE SUN)** is our Healer not Killer! Many do not know that "**SUNLIGHT**" and **Vitamin D** production is Key to staying Healthy. **Vitamin D** boosts the body's production of "**Anti-microbial Peptides**" which is a natural **anti-biotic** for **Immune protection. "Peptides are known to kill bacteria and virus"**.

When we **Qatum (Melaninite)** Beings, are in our Natural Sunny Environments like **Africa** and **South America**, places that get an ample amount of **SUNLIGHT**, then we are more Spiritually intune with **TA(EARTH)**, **MA(WATER)** and **RE(SUN)**, thus the TRUE **TAMA-REANS (Egyptians)**.

What is Qatum (Melanin) Physics?

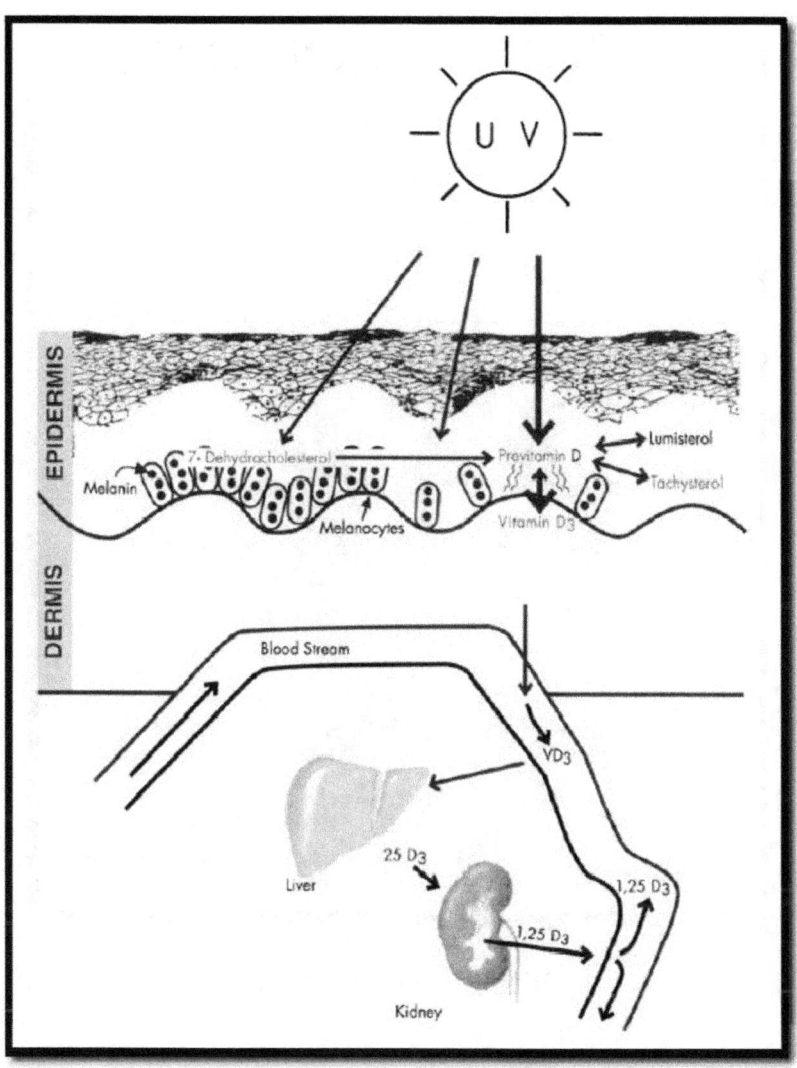

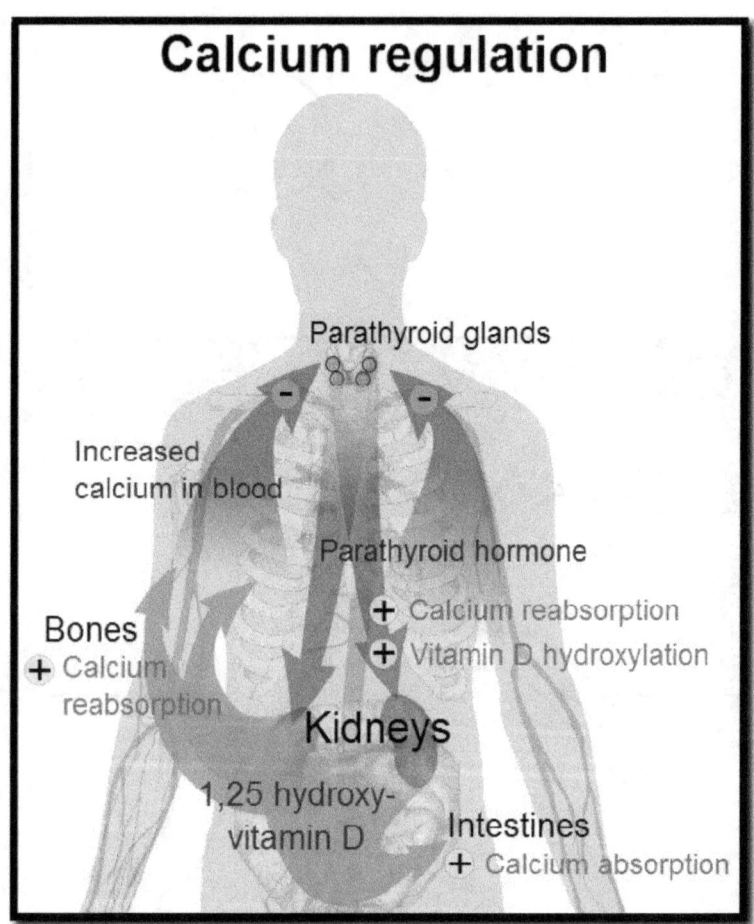

Vitamin-D Benefits!

Super Blue Green Algae and the Pineal Gland

Question: What is "Super Blue Green Algae"?

Answer: Well as you know Algae are the fundamental basis of the entire food chain - the foundational nutrient source for creating and renewing all life on earth. Because our Planet is now going through a renewal process what is being termed "The Genetic " and Dimensional Shift is taking place, and learning to ingest and eat "Primordial" foods, that tie back to the original foundation of the Planet Earth (PTAH-NUN) keeps your Mind, Body, Spirit, and Soul with the Tone, Vibration, and Frequency coming from the planet Earth (PTAH-NUN), in other words you will be able to pick up on the Telepathic signals coming from Mother Earth, and hear the messages, as to wear to be and wear to go, in order to be safe.

Question: So will taken Super Blue Green Algae open my Pineal Gland?

Answer: Not only will it open your Pineal Gland for higher transmissions coming from Earth (PTAH-NUN) and the SUN (PAA RE) but also keep the channel open! There is one such Algae that we will recommend named (AFA) **Aphanizomenon Flos-Aquae** which we call the "*Ultimate Soul Food*". **AFA**

is one of the most nutrient-rich abundant whole green foods known. This primordial blue - green alga (**Cyanobacteria**) absorbs the entire **spectrum of light** and transforms it to nutrients naturally. It contains **20 antioxidants, 68 minerals** and **70 trace elements, all amino acids** and other concentrated **photochemical**. Its abundance of **amino acids, B-vitamins** and folic acid makes **AFA the purest brain** and **nerve food. AFA supports the release of stem cells;** constituting
the **natural renewal process** of the entire body. Its **chlorophyll** content remains unmatched by any other food source including **Spirulina** and **Chlorella**. In **African countries, AFA** is sun-dried and sold at the local markets or added to sauces and poured over fish, beans or rice.

Another Interesting fact about **AFA-Super Blue Green Algae** is that it is one of your "**SMART FOODS**" or "**Brain Foods**". Let's look at it this way Intelligence is said to exist when brain cells communicate. As a whole working together Agal cells act like one Body, as a Mind/Body complex, sharing information with each other by transferring small bundles of DNA called "**Replicons**". **AFA** also contains unique low-molecular weight proteins called "**neuropeptides**". These are the basic building blocks for

Neurotransmitters which are important for proper and **optimum Brain function.**

*Note: "It is "highly" recommended if you are a person who eats a high concentration of MEAT, DIARY, STARCH what we call Mucous forming foods, to start to detox and eliminate this out your system before attempting to full open your Pineal Gland for full activation, for you will not reap the full benefits of taking in such a great Natural Food and Gift from Our Mother Nature!"

Alkaline Foods – Keeping Melanin Pure

Questions: *What other foods and Herbal Supplements I can take to keep my Melanin pure, clean and intune with the Universe?*

Answer: On the following page you will find a diagram that will get you started into knowing which foods are more Alkaline and which are more Acidic.

Alkaline = Health, Healing

Acidic = Death, Disease (Dis-ease)

What is Qatum (Melanin) Physics?

Acid | Healthy Body pH Range | Alkaline
< 5.0   5.0   5.5   6.0   6.5   7.0   7.5   8.0   8.5   9.0   9.5 +

Most Acid	Acid	Lowest Acid	FOOD CATEGORY	Lowest Alkaline	Alkaline	Most Alkaline
NutraSweet, Equal, Aspartame, Sweet 'N Low	White Sugar, Brown Sugar	Processed Honey, Molasses	SWEETENERS	Raw Honey, Raw Sugar	Maple Syrup, Rice Syrup	Stevia
Blueberries, Cranberries, Prunes	Sour Cherries, Rhubarb	Plums, Processed Fruit Juices	FRUITS	Oranges, Bananas, Cherries, Pineapple, Peaches, Avocados	Dates, Figs, Melons, Grapes, Papaya, Kiwi, Berries, Apples, Pears, Raisins	Lemons, Watermelon, Limes, Grapefruit, Mangoes, Papayas
Chocolate	Potatoes (without skins), Pinto Beans, Navy Beans, Lima Beans	Cooked Spinach, Kidney Beans, String Beans	BEANS VEGETABLES LEGUMES	Carrots, Tomatoes, Fresh Corn, Mushrooms, Cabbage, Peas, Potato Skins, Olives, Soybeans, Tofu	Okra, Squash, Green Beans, Beets, Celery, Lettuce, Zucchini, Sweet Potato, Carob	Asparagus, Onions, Vegetable Juices, Parsley, Raw Spinach, Broccoli, Garlic
Peanuts, Walnuts	Pecans, Cashews	Pumpkin Seeds, Sunflower Seeds	NUTS SEEDS	Chestnuts	Almonds	
		Corn Oil	OILS	Canola Oil	Flax Seed Oil	Olive Oil
Wheat, White Flour, Pastries, Pasta	White Rice, Corn, Buckwheat, Oats, Rye	Sprouted Wheat Bread, Spelt, Brown Rice	GRAINS CEREALS	Amaranth, Millet, Wild Rice, Quinoa		
Beef, Pork, Shellfish	Turkey, Chicken, Lamb	Venison, Cold Water Fish	MEATS			
Cheese, Homogenized Milk, Ice Cream	Raw Milk	Eggs, Butter, Yogurt, Buttermilk, Cottage Cheese	EGGS DAIRY	Soy Cheese, Soy Milk, Goat Milk, Goat Cheese, Whey	Breast Milk	
Beer, Soft Drinks	Coffee	Tea	BEVERAGES	Ginger Tea	Green Tea	Herb Teas, Lemon Water

Alkaline Foods verse Acidic Chart

224

<u>VEGETABLES</u>

Amaranth greens – same as Jamaican Callaloo, a variety of Spinach
Avocado
Katumray
Asparagus
Bell Peppers
Burro Banana
Chayote (Mexican Squash)
Cucumber
Dandelion greens
Garbanzo beans (chick peas)-optional
Izote – cactus flower/ cactus leaf- grows naturally in California
Jicama
Kale
Lettuce (all, except Iceberg)
Mushrooms (all, except Shitake)
Mustard greens
Nopales – Mexican Cactus
Okra
Olives
Onions
Poke salad -greens
Sea Vegetables (wakame/dulse/arame/hijiki/nori/kelp)
Squash
Spinach (use sparingly)
String beans
Tomato – cherry and plum only
Tomatillo
Turnip greens
Zucchini

FRUITS

Apples
Bananas – the smallest one or the Burro/mid-size (original banana) "found in most farmers markets in the U.S.A"
Berries – all varieties- Elderberries in any form – *no cranberries*
Cantaloupe
Cherries
Currants
Dates
Figs
Grapes -seeded
Limes (key limes preferred with seeds)
Mango
Melons -seeded
Orange (Seville or sour preferred, difficult to find)
Papayas
Peaches
Pears
Plums
Prunes
Raisins -seeded
Soft Jelly Coconuts
Soursops –Latin or West Indian markets)
Sugar apples (chermoya)

ALL NATURAL HERBAL TEAS

Alvaca
Anise
Chamomile
Cloves
Fennel
Ginger
Lemon grass

Red Raspberry
Sea Moss Tea

SPICES & SEASONINGS

Basil
Bay leaf
Cilantro
Dill
Marjoram
Oregano
Sweet Basil
Tarragon
Thyme
Achiote
Cayenne
Cumin
Coriander
Onion Powder
Sage

Natural Salts

Pure Sea Salt
Powdered Granulated Seaweed
Kelp/Dulce/Nori – Natural Seaweeds

Natural Sweeteners

100% Pure Maple Syrup – Grade B recommended
Maple "Sugar" (from dried maple syrup)
Date "Sugar" (from dried dates)
100% Pure Agave Syrup – (from cactus)

NUTS & SEEDS -(includes Nut & Seed Butters)

Raw Almonds and Almond butter
Raw Sesame Seeds
Raw Sesame "Tahini" Butter
Walnuts/Hazelnut

High "MAGNETIC" Alkaline Herbs

Yellow Doc
Sarsaparilla
Moringa
Pau De Arco
Nettle
Lilly of the Valley also known as "Ladder to Heaven"
Black Pepper Corn
Cloves
Black Onion Seed (Contains Melanin)

Electromagnetic "High" Alkaline Recipe

2 – 4 Lemons or Limes, Squeezed in 16oz Glass of Pure Water (**ALKALINE IS THE BEST**) with a ½ Tea Spoon of Pure Sea Salt (Preferably the rough whole SALT Grains) Eating with 2 Avocadoes and 1 cucumber! *Note: This is a "HIGHLY" Magnetic and Attractive** food recipe that will increase your **AURA**, and **Electro-Magnetic Energy**, to **attract Wealth**, Happiness and Abundance into your life!

Paa Qatum (The Melaninites)
Remember: YOU HOLD THE **MASTER KEY** TO YOUR DESTINY! IT'S ALL **WITHIN YOU** IN YOUR **DNA**!
Hotep, Peace Profound!
"GIVE CARE"

COMING SOON

COMING SOON

For information please contact: **nebheru9@gmail.com**

www.ingramcontent.com/pod-product-compliance
Lightning Source LLC
Chambersburg PA
CBHW071420180526
45170CB00001B/159